Ergebnisse der Mathematik und ihrer Grenzgebiete

Band 36

Herausgegeben von

P. R. Halmos · P. J. Hilton · R. Remmert · B. Szőkefalvi-Nagy

Unter Mitwirkung von

L. V. Ahlfors · R. Baer · F. L. Bauer · R. Courant · A. Dold
J. L. Doob · E. B. Dynkin · S. Eilenberg · M. Kneser · M. M. Postnikow
H. Rademacher · B. Segre · E. Sperner

Redaktion: P. J. Hilton

Commutation Properties
of Hilbert Space Operators
and Related Topics

C. R. Putnam

Springer-Verlag Berlin Heidelberg New York 1967

Prof. Dr. C. R. Putnam

Purdue University
Division of Mathematical Sciences
Lafayette, Indiana/U.S.A.

ISBN 978-3-642-85940-3   ISBN 978-3-642-85938-0 (eBook)
DOI 10.1007/978-3-642-85938-0

© by Springer-Verlag, Berlin · Heidelberg 1967

Softcover reprint of the hardcover 1st edition 1967

Library of Congress Catalog Card Number 66-28436

Titel-Nr. 4580

To Emogene

Preface

What could be regarded as the beginning of a theory of commutators $AB - BA$ of operators A and B on a Hilbert space, considered as a discipline in itself, goes back at least to the two papers of Weyl [3] (1928) and von Neumann [2] (1931) on quantum mechanics and the commutation relations occurring there. Here A and B were unbounded self-adjoint operators satisfying the relation $AB - BA = iI$, in some appropriate sense, and the problem was that of establishing the essential uniqueness of the pair A and B. The study of commutators of bounded operators on a Hilbert space has a more recent origin, which can probably be pinpointed as the paper of Wintner [6] (1947). An investigation of a few related topics in the subject is the main concern of this brief monograph.

The ensuing work considers commuting or "almost" commuting quantities A and B, usually bounded or unbounded operators on a Hilbert space, but occasionally regarded as elements of some normed space. An attempt is made to stress the role of the commutator $AB - BA$, and to investigate its properties, as well as those of its components A and B when the latter are subject to various restrictions. Some applications of the results obtained are made to quantum mechanics, perturbation theory, Laurent and Toeplitz operators, singular integral transformations, and Jacobi matrices.

In the greater part of Chapter I, A and B are bounded operators on a Hilbert space, although the boundedness restriction is relaxed in § 1.6, where, incidentally, spectral resolution theory is first used. Chapter II is concerned mostly with commutators when one of the components is self-adjoint and is bounded or at least half-bounded, and where the commutator itself is self-adjoint and non-negative. This situation is related to that dealing with the unitary equivalence of self-adjoint operators the difference of which is non-negative. Spectral resolution theory and the concept of absolutely continuous spectrum play an important role here. Chapter III deals with bounded operators T which are semi-normal, that is, which are such that $TT^* - T^*T$ is semi-definite. The spectra of T and its real and imaginary parts are studied. Chapter IV deals with the commutation, and to a lesser extent, with the anti-commutation relations occurring in quantum physics. The operators involved are for the most part unbounded and self-adjoint and most of the results deal with the problem of the essential uniqueness of the operators A and B satisfying $AB - BA = iI$. Chapter V considers the issue of the unitary equivalence of a self-adjoint operator and a suitable perturbed form of this operator. A foretaste of some of this occurs earlier in Chapter II wherein are derived some results used near the end of Chapter V. The problems in Chapter V, as in the preceding chapter, are regarded in a mathematical light but are largely motivated by a branch of quantum physics, in this case, that of scattering theory. Finally, Chapter VI is devoted to a few applications of the theory developed earlier. The treatment of the topics considered—Laurent and Toeplitz operators, certain singular integral operators, Jacobi matrices—is not exhaustive and is mainly intended to point up the role and potential of commutator theory in these fields.

Needless to say, the topics included form only a small portion of the possible array of subjects comprehended by the title. In particular only brief attention is given to the rather extensive commutation theory dealing with the case where the underlying Hilbert space is exclusively finite-dimensional, and which is largely dependent on special concepts (elementary divisors, normal forms, etc.) for the operators involved.

In most instances complete proofs are given, although they are occasionally only outlined or omitted entirely.

Each reference in the bibliography has been alluded to specifically in the body of the text, sometimes only mentioned however. Although no completeness is intended or claimed for the bibliography, it is hoped that it is dense, in some appropriate metric, and can supply an investigator with ample leads.

It is a sincere pleasure to thank Professor B. Sz.-Nagy for his kind invitation to me to write a monograph in the Ergebnisse series and also for his advice and help in the preparation of this work and for his continued interest in its progress. In addition, acknowledgment is hereby gratefully extended to the National Science Foundation for its financial support, in part, of the project, and to the Springer-Verlag for its obviously indispensable role in the entire undertaking.

West Lafayette, Indiana 1966   C. R. Putnam

Contents

Contents

Chapter I

Commutators of bounded operators

§ 1.1 Introduction

A few general references for this work are Akhiezer and Glazman [1], Berberian [2], Dunford and Schwartz [1], Halmos [3,7], Naimark [1], von Neumann [7], Riesz and Sz.-Nagy [1], Stone [1], Sz.-Nagy [1, 6], Taylor [2], Wintner [2], Zaanen [1].

In the sequel, $\mathfrak{H}$ will denote a Hilbert space (complete, complex inner product space) of elements $x, y, \ldots$, with inner product (x, y) and norm $\|x\| = (x, x)^{\frac{1}{2}}$. A transformation A on $\mathfrak{H}$ is said to be bounded if it is linear, its domain is the entire space $\mathfrak{H}$, and if there exists a finite constant $M \geq 0$ for which $\|Ax\| \leq M \|x\|$ for all x in $\mathfrak{H}$. The norm $\|A\|$ of A is defined by $\|A\| = \inf M$ or, equivalently, by $\|A\| = \sup \|Ax\|$ with $\|x\| = 1$.

In large part this chapter will be concerned with commutators

$$AB - BA = C \qquad (1.1.1)$$

of bounded operators A and B on a (non-trivial) Hilbert space $\mathfrak{H}$, usually of infinite dimension.

§ 1.2 Structure of commutators of bounded operators

One can ask what bounded operators C on a Hilbert space $\mathfrak{H}$ can be expressed in the form of a commutator (1.1.1), where A and B are bounded operators on $\mathfrak{H}$.

In the special case in which $\mathfrak{H}$ has finite dimension n, it can be supposed that A and B are finite matrices acting on the space of n-component vectors. Since the trace, $\text{tr}(A) = \Sigma a_{kk}$, of a matrix $A = (a_{ij})$ is linear and satisfies $\text{tr}(AB) = \text{tr}(BA)$, a necessary condition that C be a commutator is that its trace be zero. It turns out that this condition is also sufficient; see Shoda [1], where the underlying field need not be that of the complex numbers but any one of characteristic zero, and Albert and Muckenhoupt [1], for fields of arbitrary characteristic. For a related result, see also Thompson [1].

In case $\mathfrak{H}$ is infinite-dimensional the problem has only recently been solved. In order to discuss the issue from its beginning consider first the following result of Wintner [6].

Theorem 1.2.1. *The identity operator on a Hilbert space cannot be expressed as the commutator of bounded operators.*

Proof. It will be convenient to recall first a few properties of operators and their spectra. A bounded operator A is said to be non-singular if it has a bounded (unique, two-sided) inverse A^{-1}; otherwise, A is said to be singular. The spectrum, $\mathrm{sp}(A)$, of A is the set of complex numbers λ for which $A - \lambda I$ is singular. The spectrum is a non-empty, closed, bounded subset of the complex plane (Wintner [1]). In case A is non-singular then $AB - \lambda I = A(BA - \lambda I)A^{-1}$, so that AB and BA have identical spectra.

Suppose that the assertion of the theorem is false, so that $AB - BA = I$ holds for some pair of bounded operators A and B. Since the equation continues to hold if A is replaced by $A - \lambda I$, and since λ can be chosen so as not to belong to the spectrum of A, say $|\lambda| > \|A\|$, there is no loss of generality in supposing that A is non-singular and hence $\mathrm{sp}(AB) = \mathrm{sp}(BA)$. Since AB is bounded, its spectrum contains at least one point, say μ. It then follows from $AB - BA = I$ that $\mu + 1$ also belongs to $\mathrm{sp}(AB)$. A repetition of the argument shows that $\mu + n$ must belong to $\mathrm{sp}(AB)$ for each $n = 0, \pm 1, \pm 2, \ldots$, in contradiction with the fact that $\mathrm{sp}(AB)$ must be bounded. This completes the proof.

Remark. In view of the quantum mechanical motivation of the above theorem as first given by Wintner [6], he supposed that A and B were self-adjoint. His proof is valid without this hypothesis however and is essentially as given above. It may be noted that his paper has often been overlooked by mathematicians, a circumstance probably in large part attributable to its publication in a physical, rather than a mathematical, journal.

Shortly after the appearance of Wintner's result, Wielandt [1] proved the following generalization.

Theorem 1.2.2. *In an arbitrary normed algebra with a unit element I, the element I cannot be expressed as the commutator of elements in the algebra.*

Proof. Suppose the contrary, that is, suppose that $AB - BA = I$. An induction argument shows that $(n+1)B^n = AB^{n+1} - B^{n+1}A$ holds for $n = 0, 1, 2, \ldots$ Hence $(n+1)\|B^n\| \leq 2\|A\| \|B^{n+1}\| \leq 2\|A\| \|B\| \|B^n\|$, and so $B^n = 0$ for n sufficiently large. But it is clear that $B^N = 0$ for some N implies $0 = B^N = B^{N-1} = \ldots = B^0 = I$, a contradiction, and the proof is complete.

For a generalization of the Wintner-Wielandt result see Singer and Wermer [1]. Another is the following due to Halmos [9].

Theorem 1.2.3. *Let $\mathfrak{H}$ be an infinite-dimensional Hilbert space. If C denotes any completely continuous operator on $\mathfrak{H}$ then $I + C$ is not a commutator.*

Proof. It is known that the set of all bounded operators A on a Hilbert space $\mathfrak{H}$, with norm $\|A\|$, constitutes a Banach algebra $B = B(\mathfrak{H})$. Moreover, the set C of completely continuous operators forms a closed two-sided ideal in B. The quotient space $B' = B/C$ consists of elements A', each of which is a class of elements in B equivalent to a fixed element A in B modulo C. That is, $A' = \{A + C\}$ where A is fixed in B and C runs through C. It is known that B' with norm given by $\|A'\| = \inf\|X\|$, X in A', is a Banach algebra (cf. Calkin [1], Naimark [1], pp. 73, 178, Schatten [1]).

Since $\mathfrak{H}$ is infinite-dimensional, I is not in C and B' is not the trivial space. According to the Wintner-Wielandt result, the identity element I' of B' is not a commutator, that is, in terms of the space B, $I + C$ is not a commutator whenever C is completely continuous. This completes the proof.

So far, it has been shown that on an infinite-dimensional Hilbert space, if $D = \lambda I + C$ where $\lambda \neq 0$ and C is completely continuous, then D is not a commutator. The following result of Brown and Pearcy [1, 2] is the converse assertion, at least if the space is separable, and completes the cycle.

Theorem 1.2.4. *If $\mathfrak{H}$ is an infinite-dimensional separable Hilbert space and if D is not a commutator on $\mathfrak{H}$ then D is of the form $D = \lambda I + C$ where $\lambda \neq 0$ and C is completely continuous.*

Proof. (Outline). The proof is rather lengthy and only a brief indication will be given. That every completely continuous operator on $\mathfrak{H}$ is a commutator was proved by Brown, Halmos and Pearcy [1]. Hence it is sufficient to show that every element of the class of operators not of the form $\lambda I + C$ with λ arbitrary and C completely continuous, is a commutator. It can be shown that each such element F is similar to a

3×3 operator matrix $\begin{pmatrix} A_{11} & A_{12} & 0 \\ A_{21} & A_{22} & I \\ A_{31} & A_{32} & 0 \end{pmatrix}$ acting on the direct sum of three

copies of $\mathfrak{H}$. From this it can be deduced that F is similar to an operator on the direct sum of two copies of $\mathfrak{H}$ and which is of the form $\begin{pmatrix} A & V \\ B & 0 \end{pmatrix}$, where V is isometric with the property that the null space of V^* is infinite-dimensional. The proof of the theorem is then completed by showing that the latter matrix is a commutator. A key role is played by the theorem of Pearcy [1] that a bounded operator having a "large" (that is, in the present case, infinite-dimensional) null space is a commutator, a result obtained by generalization of a construction of Halmos [5]. Finally, it may be noted that use also is made of a result of Lumer and Rosenblum [1].

Remark. Brown and Pearcy [2] also obtain a result analogous to Theorem 1.2.4 in case $\mathfrak{H}$ is non-separable. In addition they discuss possible generalizations of their methods and results in connection with Theorem 1.2.4 to C^* algebras and von Neumann algebras. In particular, using the fact that $B(\mathfrak{H})/C$ is a C^* algebra in case $\mathfrak{H}$ is separable (Calkin [1]) they note that the result of Theorem 1.2.4 implies the existence of a non-trivial C^* algebra of operators with a unit having the property that all non-scalar elements are commutators.

It can be noted that Theorem 1.2.4 answers a number of questions concerning commutators raised by Halmos [4,5,9].

As has been shown above, if $\mathfrak{H}$ is an infinite-dimensional separable Hilbert space, then a bounded operator A on $\mathfrak{H}$ is a commutator if and only if A is not of the form $A = \lambda I + C$ where $\lambda \neq 0$ and C belongs to the ideal, C, of completely continuous operators. (In case the space is non-separable an analogous assertion holds but with C replaced by a certain other ideal playing a role similar to that of C in the separable case; see Brown and Pearcy [2].) As has been noted by Brown and Pearcy [2], this result implies that the set of all commutators on $\mathfrak{H}$ is dense in the uniform norm topology, so that, in particular, the identity operator, although not a commutator (Theorems 1.2.1 and 1.2.2) is the limit in the norm of commutators. In addition one also obtains easily the following interesting result of Halmos [5].

Theorem 1.2.5. *Every bounded operator on an infinite-dimensional Hilbert space is the sum of two commutators.*

In line with Kaplansky's observation as mentioned by Halmos [4], where the assertion of Theorem 1.2.5 was proved with the "two" replaced by "four," it can be noted here that Theorem 1.2.5 implies in particular that an additive functional defined on $B(\mathfrak{H})$ and equal to zero on the commutators must be identically zero, so that it is impossible to extend the idea of "trace" to operators on infinite-dimensional Hilbert spaces.

§ 1.3 Commutators $C = AB - BA$ with $AC = CA$

Jacobson ([1], Lemma 2) proved that, in an associative algebra of finite rank over a field of characteristic zero, if $C = AB - BA$ commutes with A, then C is nilpotent. The following analogous result for normed algebras was conjectured by Kaplansky (cf. Halmos [5], .p. 192) and proved independently by Kleinecke [1] and Sirokov [1].

Theorem 1.3.1. *Let A and B be elements of a normed algebra and let $C = AB - BA$ commute with A, that is,*

$$AC = CA . \tag{1.3.1}$$

Then C is generalized nilpotent, that is,

$$\lim_{n \to \infty} \| C^n \|^{1/n} = 0 . \tag{1.3.2}$$

An immediate consequence is the following.

Corollary. *If A and B are bounded operators on a Hilbert space and if C = AB − BA satisfies* (1.3.1) *then the spectrum of C is the single number* 0.

Special cases of the Corollary, in which it was assumed also that *B* commutes with *C*, were proved earlier by Putnam [8] and, for any Banach algebra, by Vidav [1]; see also Singer and Wermer [1], p. 264.

Kleinecke [1] has observed that Theorem 1.3.1 is a consequence of the following more general theorem.

Theorem 1.3.2. *Let Δ denote a differentiation operator on a ring, so that $\Delta(AB) = (\Delta A)B + A(\Delta B)$ and let $\Delta^2 T = 0$. Then*

$$\Delta^n(T^n) = n!(\Delta T)^n .\qquad(1.3.3)$$

Proof of Theorem 1.3.1 granting Theorem 1.3.2. If $\Delta X = AX - XA$ then it is easily seen that Δ acts as a differentiation on the algebra. Furthermore the assumption (1.3.1) implies that $\Delta^2 B = 0$. An application of Theorem 1.3.2 then yields $C^n = (\Delta B)^n = \Delta^n(B^n)/n!$. But $\|\Delta X\| = \|AX - XA\| \leq 2\|A\|\,\|X\|$ and so Δ is a bounded operator with norm not exceeding $2\|A\|$. Hence $\|C^n\| \leq \|\Delta^n\|\,\|B\|^n/n! \leq 2^n\|A\|^n\|B\|^n/n!$ and so (1.3.2) follows.

Proof of Theorem 1.3.2. The proof follows by an induction argument. That (1.3.3) holds if $n = 1$ is clear. If one assumes $\Delta^{n-1}(T^{n-1}) = (n-1)! \times (\Delta T)^{n-1}$ then differentiation and the use of $\Delta^2 T = 0$ yield $\Delta^n(T^{n-1}) = 0$. Furthermore, by Leibnitz's rule,

$$\Delta^n(T^n) = \Delta^n(TT^{n-1}) = \sum_{k=0}^{n}\binom{n}{k}\Delta^k T\Delta^{n-k}(T^{n-1})$$

$$= T\Delta^n(T^{n-1}) + n\Delta T\Delta^{n-1}(T^{n-1})$$

$$= 0 + n\Delta T(n-1)!\,(\Delta T)^{n-1} = n!(\Delta T)^n .$$

This completes the proof.

If *A* denotes a bounded operator on a Hilbert space then $C = AA^* - A^*A$ is self-adjoint. Hence, Theorem 1.3.1 implies that if *A* commutes with AA^* and A^*A then $AA^* = A^*A$, that is, *A* is normal (Brown [1]) or that if only *A* commutes with $AA^* - A^*A$ then *A* is normal (Putnam [8]).

For some related material see the survey article of Kaplansky [2], pp. 20–21. See also Bellman [1], Kato and Taussky [1], Kuzmin [1], McCoy [1], Marcus and Khan [1], Sakai [1, 2].

§ 1.4 Multiplicative commutators

The following result of Putnam and Wintner [2] is a kind of analogue of Theorem 1.3.1.

Theorem 1.4.1. *Let A denote a normed algebra with a unit I, let A and B be invertible elements of A, and suppose that A commutes with $C = AB - BA$. In addition, suppose that A has a logarithm E which commutes with every X in A which commutes with A, thus*

$$A = e^E, \qquad AX = XA \text{ implies } EX = XE. \tag{1.4.1}$$

Then if D denotes the multiplicative commutator

$$D = ABA^{-1}B^{-1}, \tag{1.4.2}$$

the element $D - I$ is generalized nilpotent, that is,

$$\lim_{n \to \infty} \|(D - I)^n\|^{1/n} = 0 \quad (\text{that is, } \mathrm{sp}(D) = 1 \text{ only}). \tag{1.4.3}$$

Remark. Recall that the spectrum of an element A in A is defined in a manner similar to that for a bounded operator on a Hilbert space $\mathfrak{H}$, these latter elements constituting the Banach algebra $B(\mathfrak{H})$; see § 1.2 above. Thus the spectrum of A is the complement of the set of complex numbers λ for which $(A - \lambda I)^{-1}$ is in A. That the spectrum is non-empty is customarily proved by a modification of Wintner's original argument (Wintner [1]), for the special case when $A = B(\mathfrak{H})$.

Proof. The assumption $AC = CA$ implies readily that $e^{tA} B e^{-tA} = B + tC$ for all complex t; see Campbell [1], pp. 385–386, Hausdorff [1], p. 26, also Birkhoff [1]. Since the left side is non-singular for all t and since $B + tC = (I + tCB^{-1})B$, then $I + tCB^{-1}$ is non-singular for all t and hence CB^{-1} is generalized nilpotent. Since $EA = AE$, an easy calculation shows that $AF - FA = EC - CE$ where $F = EB - BE$. Since C commutes with A it commutes also with E (by hypothesis), hence $AF = FA$ and so $EF = FE$. Hence (see above) $e^{tE} B e^{-tE} = B + tF = (I + tFB^{-1})B$ and FB^{-1} is generalized nilpotent. Also, if $t = 1$, $e^E B e^{-E} = B + F$, that is, $AB - BA = FA$ or $C = FA$. Hence $CA^{-1}B^{-1}$ $(= FB^{-1})$ is generalized nilpotent, that is, (1.4.3), and the proof is complete.

Let A be a bounded operator on $\mathfrak{H}$. Then its spectrum is bounded and closed, hence the canonical decomposition of its open complement as a union of connected open sets contains a unique unbounded component. It was shown by Wintner [8] that if 0 belongs to this component then A has a logarithm L; further, it is clear from his proof that such a logarithm L exists which can be expressed as a power series in A. In particular, L commutes with any operator commuting with A. In view of these remarks, Theorem 1.4.1 has the following consequence (Putnam and Wintner [2]).

Corollary. *Let A and B be non-singular elements of the Banach algebra $B(\mathfrak{H})$ and suppose that A commutes with $AB - BA$. In addition suppose that 0 belongs to the unbounded component of the complement of $\mathrm{sp}(A)$. Then (1.4.3) holds, where D is defined by (1.4.2).*

If A is non-singular the requirement on sp(A) in the Corollary is certainly fulfilled if the space $\mathfrak{H}$ is finite-dimensional, and also in the infinite-dimensional case, if, for instance, A differs from a completely continuous operator by a (non-zero) multiple of the identity.

It is an open question whether either Theorem 1.4.1 or the Corollary holds if the assumption concerning the logarithm of A is relaxed to either of the requirements (i) that A have some logarithm or, to (ii) that A be only non-singular. As to (i) simple examples even in the finite-dimensional case show that a non-singular operator A can have logarithms not satisfying (1.4.1). For instance, if A is the 2×2 unit matrix, such a logarithm is given by $E = \begin{pmatrix} 0 & 0 \\ 0 & 2\pi i \end{pmatrix}$. The unit matrix of course also possesses the logarithm $E = 0$ which does satisfy (1.4.1). As for (ii), it can be noted that Halmos, Lumer and Schäffer [1] (cf. also Schäffer [1]) showed that there exist non-singular operators on infinite-dimensional spaces which not only fail to possess any logarithms but do not even have square roots. See also Halmos and Lumer [1] and Deckard and Pearcy [1].

In line with Wintner's remarks in his paper [8], it can be noted that his argument there yields a simple proof of the well-known result in the finite-dimensional case that a non-singular matrix has a logarithm, without the somewhat cumbersome theory of canonical forms for matrices.

Concerning logarithms of elements of a Banach algebra see also Hille [1], Krabbe [1].

The following result is due to Herstein [1].

Theorem 1.4.2. *Let A and B be invertible elements of a Banach algebra and suppose that A commutes with $C = AB - BA$. Then whenever λ is in the spectrum of D of (1.4.2), so also is $2 - 1/\lambda$.*

Proof. Since $D - I = CA^{-1} B^{-1} = A^{-1} CB^{-1}$, then $I - D^{-1} = CB^{-1} A^{-1} = A(D - I)A^{-1}$, from which the assertion follows.

Herstein [1] also gives a simple algebraic proof of an analogue of Theorem 1.4.1 for $n \times n$ matrices over a field with characteristic greater than n.

Concerning the general problem as to what elements of a normed algebra or, in particular, what bounded operators on a Hilbert space $\mathfrak{H}$, are multiplicative commutators, that is, are of the form (1.4.2), the following can be mentioned. In case $\mathfrak{H}$ is finite-dimensional, it was shown by Shoda [2] that an operator D on $\mathfrak{H}$ is a multiplicative commutator if and only if D has determinant 1. Recently, Brown and Pearcy [3] have investigated the corresponding problem in the infinite-dimensional case by applying methods similar to those they used in their paper [2] for additive commutators. They completely settle the problem in the special case of normal operators D and have established, among other things, the following two theorems, which will be stated without proof.

Theorem 1.4.3. *A non-singular normal operator D on an infinite-dimensional separable Hilbert space $\mathfrak{H}$ is a multiplicative commutator if and only if D is not of the form $D = \lambda I + Q$, where $|\lambda| \neq 1$ and Q is completely continuous.*

In particular, every unitary operator on such a Hilbert space is a multiplicative commutator.

Theorem 1.4.4. *Every non-singular operator on an infinite-dimensional separable Hilbert space is the product of two multiplicative commutators.*

For some other results on multiplicative commutators see Pasiencier and Wang [1], Putnam [19,20], Robinson [1], Taussky [1,2].

§ 1.5 Commutators and numerical range

If A is a bounded operator on a Hilbert space $\mathfrak{H}$ of elements x, the set of complex numbers (Ax, x) where $\|x\| = 1$ is called the numerical range of A. The set is convex (Hausdorff-Toeplitz); see, e.g., Stone [1], p. 131, Wintner [2], p. 34, also Donoghue [1], Halmos [10]. If W_A denotes the closure of the above convex set, then W_A always contains the spectrum of A. Furthermore, if A is normal, W_A is the least closed convex set containing the spectrum of A (Toeplitz).

Wintner raised the question (cf. Putnam [1], p. 127) whether the number 0 belongs to the set W_C if C is a commutator $C = AB - BA$ of bounded operators A and B. That the answer is yes in case $\mathfrak{H}$ is finite-dimensional can be deduced from the following simple argument. If $C = (c_{ij})$ then $c_{11} = (Cx, x)$ for $x = (1, 0, 0, \ldots)$. Similarly each diagonal element c_{kk} of C is in W_C. Since W_C is convex the center of mass of the system of n unit masses situated at the points c_{kk}, that is $\operatorname{tr}(C)/n$, belongs to W_C. But $\operatorname{tr}(C) = 0$ and so 0 is in W_C.

That Wintner's question has a negative answer for arbitrary A and B on an arbitrary Hilbert space $\mathfrak{H}$ was first proved by Halmos [4]. This result can be concluded from the results of § 1.2 by choosing C not of the form $\lambda I + D$ with $\lambda \neq 0$ and D completely continuous and so that 0 is not in W_C.

On the other hand, if one of the components of C, say A, is normal ($AA^* = A^*A$) or even semi-normal ($AA^* - A^*A$ semi-definite), the question is answered affirmatively by the following result of Putnam [1].

Theorem 1.5.1. *Let A and B denote bounded operators on a Hilbert space $\mathfrak{H}$ and suppose that $AA^* - A^*A$ is semi-definite. Then 0 belongs to the set W_C, where $C = AB - BA$.*

Proof. Let λ and θ, with $\lambda \neq 0$, be real numbers and define the bounded operator $F = F(\lambda, \theta) = |\lambda|^{\frac{1}{2}} e^{i\theta} A + |\lambda|^{-\frac{1}{2}} B^*$. A simple calculation shows that

$$\operatorname{sgn} \lambda (FF^* - F^*F) = D(\lambda) + (\operatorname{sgn} \lambda) H(\theta), \qquad (1.5.1)$$

where

$$H(\theta) = e^{i\theta}C + e^{-i\theta}C^* \quad \text{and} \quad D(\lambda) = \lambda(AA^* - A^*A) - \lambda^{-1}(BB^* - B^*B).$$

It will first be shown that if

$$\tau = \inf_{\|x\|=1} |(Cx, x)|,$$

then

$$\eta \geq 0 \quad \text{and} \quad \eta \geq \inf_{\lambda \neq 0} (\max D(\lambda)) \text{ implies } \tau \leq \tfrac{1}{2}\eta. \tag{1.5.2}$$

(Here, if T is any self-adjoint operator, max T and min T will denote the maximum and minimum points of the spectrum of T.) If (1.5.2) were false then there would exist some $\eta \geq 0$ satisfying $\eta \geq \inf_{\lambda \neq 0} (\max D(\lambda))$ and $\tau > \tfrac{1}{2}\eta$. Since $\tau > 0$ and since $W_C e^{i\theta}$ is the set W_C rotated by θ it is possible to choose values θ so that either $\text{Re}(e^{i\theta} W_C) \geq \tau$ or $\text{Re}(e^{i\theta} W_C) \leq -\tau$. Clearly there exists some value of λ, say $\lambda = \mu$ ($\neq 0$), such that $\max D(\mu) < 2\tau$. Then choose $\theta = \theta(\mu)$ so that $(\text{sgn } \mu) \text{Re}(e^{i\theta} W_C) \leq -\tau$. Clearly $D(\mu) + (\text{sgn } \mu) H(\theta)$ is negative definite and, by (1.5.1), the operator $FF^* - F^*F \equiv G$ with $F = F(\mu, \theta(\mu))$, is definite. But this is impossible. For, note that $G = F_\lambda F_\lambda^* - F_\lambda^* F_\lambda$ with $F_\lambda = F - \lambda I$ and then choose λ so that F_λ is non-singular. Then $\text{sp}(F_\lambda^* F_\lambda) = \text{sp}(F_\lambda^* F_\lambda)$ (see the preliminary remarks occurring in the proof of Theorem 1.2.1). Consequently, if $G > 0$ then $\max F_\lambda F_\lambda^* \geq \max F_\lambda^* F_\lambda + \min G$, yielding a contradiction. Similarly $G < 0$ is impossible. This completes the proof of the implication (1.5.2).

The assertion of Theorem 1.5.1 now can be deduced from (1.5.2). Clearly there is no loss of generality in supposing that $AA^* - A^*A \leq 0$. Then, for $\lambda > 0$, the definition of $D(\lambda)$ shows that $D(\lambda) \leq -\lambda^{-1}(BB^* - B^*B) \leq 2\lambda^{-1} \|B\|^2 I$ and hence $\max D(\lambda) \leq 2\lambda^{-1} \|B\|^2$ for all $\lambda > 0$. This means that it is possible to choose $\eta = 0$ in (1.5.2), and hence $\tau \equiv \inf_{\|x\|=1} |(Cx, x)| = 0$, as was to be shown. This completes the proof of Theorem 1.5.1.

For a related result, see Putnam [25].

§ 1.6 Some results on normal operators

Von Neumann [6] (pp. 60–62) raised the question as to whether $AB = BA$ with A normal and B arbitrary (both bounded) implies $A^*B = BA^*$. The problem was answered affirmatively by Fuglede [1] who proved the following.

Theorem 1.6.1. *If B is any bounded operator, if A is normal but not necessarily bounded, and if $BA \subset AB$ then $BA^* \subset A^*B$.*

For the basic properties of operators needed here, see, e.g., Sz.-Nagy [1].

Proof. The proof to be given is that of Fuglede and depends on the spectral resolution formula $A = \int z \, dK_z$. If α_1 and α_2 denote disjoint Borel sets of the complex plane, it will be shown that

$$Q \equiv K(\alpha_1) B K(\alpha_2) = 0 . \qquad (1.6.1)$$

(Here $K(\alpha)$ denotes the projection operator associated with the Borel set α by the spectral family K_z; cf. e.g., Halmos [3].) First, suppose that α_1 and α_2 are bounded and at a positive distance d from one another. Since α_2 is bounded then for any x in $\mathfrak{D}_A$,

$$K(\alpha_2) x \in \mathfrak{D}_A \quad (= \mathfrak{D}_{BA})$$

and so

$$B \int_{\alpha_2} z \, dK_z x = ABK(\alpha_2) x .$$

Hence, on applying the operator $K(\alpha_1)$ and noting that $\mathfrak{D}_A$ is dense, one obtains

$$K(\alpha_1) B \int_{\alpha_2} z \, dK_z = \int_{\alpha_1} z \, dK_z BK(\alpha_2) .$$

If z_1 and z_2 are arbitrary numbers in α_1 and α_2 respectively, this can be written as

$$\int_{\alpha_1} (z - z_1) dK_z Q = Q \int_{\alpha_2} (z - z_2) dK_z + (z_2 - z_1) Q .$$

Hence $|z_2 - z_1| \, \|Q\| \leq 2\delta \|Q\|$, where δ denotes the maximum diameter of α_1 and α_2, and so $Q = 0$ whenever $\delta < \frac{1}{2} d$. But this implies that (1.6.1) holds for arbitrary bounded disjoint α_1 and α_2 provided $d > 0$. For one need only express each $\alpha_k (k = 1, 2)$ as a finite union $\alpha_k = \Sigma_j \alpha_{kj}$ of disjoint Borel sets α_{kj}, the diameter of each of which is less than $\frac{1}{2} d$, and then apply the result already proved.

Next, relation (1.6.1) holds whenever α_1 and α_2 are bounded and disjoint. This can be readily deduced by expressing $K(\alpha_1)$ as the strong limit of a sequence of projections $K(\alpha_1^n)$, where α_1^n is at a positive distance from α_2. Finally it is clear that (1.6.1) holds even if α_1 and α_2 are unbounded, provided that their intersection is empty. Thus the relation (1.6.1) is established in general.

Now let α denote any Borel set. Then (1.6.1) implies that

$$K(\alpha) B = K(\alpha) BI = K(\alpha) B(K(\alpha) + K(\alpha')) = K(\alpha) BK(\alpha) ,$$

where α' denotes the complement of α. Similarly $BK(\alpha) = K(\alpha) BK(\alpha)$ and so $K(\alpha) B = BK(\alpha)$. This implies in particular that $BA^* \subset A^* B$ and the proof is complete.

A generalization of Theorem 1.6.1 due to Putnam [2] is the following.

Theorem 1.6.2. *If B is bounded, if A_1 and A_2 are normal and if $BA_1 \subset A_2 B$, then $BA_1^* \subset A_2^* B$.*

Proof. A proof can be given along the lines of Fuglede's proof of Theorem 1.6.1 given above. For if $A_j = \int z \, dK_{jz}$ $(j=1, 2)$, a similar argument shows first that $K_2(\alpha_2) B K_1(\alpha_1) = 0$ whenever α_1 and α_2 are disjoint Borel sets of the complex plane. This then implies $K_2(\alpha) B = B K_1(\alpha)$ for any Borel set α and the proof is complete.

Other proofs of Theorem 1.6.1 have been given by Halmos [1, 3] and Dunford [1], who has generalized Fuglede's theorem to the case where A is not necessarily normal but is, in his theory, spectral. Below is given another proof of Theorem 1.6.2 due to Rosenblum [5].

Proof. Suppose first that A_1 and A_2 are bounded and that $BA_1 = A_2 B$. Then

$$BA_1^k = A_2^k B \quad \text{for} \quad k = 0, 1, 2, \ldots,$$

and hence

$$B \exp(itA_1) = \exp(itA_2) B \quad \text{for all complex } t.$$

Hence $B = \exp(itA_2) B \exp(-itA_1)$

and

$$\exp(itA_2^*) B \exp(-itA_1^*)$$

$$= \exp(itA_2^*) \exp(itA_2) B \exp(-itA_1) \exp(-itA_1^*)$$

$$= \exp(itA_2^* + itA_2) B \exp(-itA_1 - itA_1^*),$$

the last equality by virtue of the normality of A_1 and A_2. Since $tA_2^* + tA_2$ and $tA_1 + tA_1^*$ are self-adjoint, the last two exponential operators are unitary and therefore of norm 1. Hence $\exp(itA_2^*) B \exp(-itA_1^*)$ is bounded and analytic for all t and, by Liouville's theorem (see Hille and Phillips [1]), must be a constant. Thus $\exp(itA_2^*) B \exp(-itA_1^*) = B$, so that a differentiation with respect to t followed by setting $t=0$ implies $A_2^* B = B A_1^*$.

Next, let A_1 and A_2, with spectral resolutions $A_j = \int z \, dK_{jz}$, be possibly unbounded. If x is in the domain of A_1, then $BA_1 \subset A_2 B$ implies that Bx is in the domain of A_2. Also, both operators A_j and A_j^* (for each fixed $j=1$ and 2) have identical domains. If $\mu > 0$ and α_μ denotes the closed disk $|z| \leqq \mu$, then $A_1 K_1(\alpha_\mu)$ and $A_2 K_2(\alpha_\nu)$ $(\mu, \nu > 0)$ are bounded normal operators and $R A_1 K_1(\alpha_\mu) = A_2 K_2(\alpha_\nu) R$, where $R = K_2(\alpha_\nu) B K_1(\alpha_\mu)$. By the result already proved for bounded operators, $R[A_1 K_1(\alpha_\mu)]^* = [A_2 K_2(\alpha_\nu)]^* R$, that is, $K_2(\alpha_\nu) B A_1^* K_1(\alpha_\mu) x = K_2(\alpha_\nu) A_2^* B K_1(\alpha_\mu) x$, for all x in the domain of A_1^*. On taking the strong limits first $\mu \to \infty$ and then $\nu \to \infty$, it follows that $B A_1^* x = A_2^* B x$ and so $B A_1^* \subset A_2^* B$ as was to be proved.

Remark. That the assertion of Theorem 1.6.2 becomes false in case one of the operators A_1, A_2 is normal while the other is only semi-normal is shown by an example of Stampfli [1], p. 1456.

The next theorem was given in Putnam [2] and will be derived as a consequence of Theorem 1.6.1.

Theorem 1.6.3. *Let A be normal, B and C be bounded, and suppose that $BA + C \subset AB$ and $CA \subset AC$. Then $C = 0$, that is, $BA \subset AB$.*

Proof. Let α_1 and α_2 denote disjoint Borel sets of the complex plane. As before, if $A = \int z \, dK_z$, it follows from Theorem 1.6.1 that $K(\alpha_1) CK(\alpha_2) = CK(\alpha_1) K(\alpha_2) = 0$. Consequently, the type of argument used earlier now leads to $K(\alpha) B = BK(\alpha)$ for any Borel set α, hence $C = 0$, and the proof is complete.

For a discussion of some generalizations of Theorem 1.6.3, see Kaplansky [2], pp. 20–21.

The following was also proved in Putnam [2] as a consequence of Fuglede's theorem. (See also Berberian [1].)

Theorem 1.6.4. *Let A_1 and A_2 denote similar bounded normal operators, so that $A_2 = BA_1 B^{-1}$ holds for some bounded non-singular B. If $B = PU$ is the polar factorization of B, where P is positive definite and U is unitary then $A_2 = UA_1 U^*$. Thus, similarity equivalence of normal operators implies their unitary equivalence.*

Proof. Since $BA_1 = A_2 B$ then Fuglede's theorem yields $BA_1^* = A_2^* B$ and, on taking adjoints, $A_1 B^* = B^* A_2$. Hence $BB^* A_2 = BA_1 B^* = A_2 BB^*$. Since $BB^* = P^2$ then A_2 commutes with P^2 and, since P is positive, A_2 commutes with P. This implies that $PUA_1 = A_2 PU = PA_2 U$ and hence $UA_1 = A_2 U$ as was to be proved.

Corollary. *Let A_1 and A_2 be arbitrary (not necessarily bounded) normal operators and suppose that B is a non-singular bounded operator for which $BA_1 \subset A_2 B$. If, as above, $B = PU$ and if $A_j = \int z \, dK_{jz}$ then again $K_{2z} = UK_{1z} U^*$.*

Proof. According to Theorem 1.6.2, the assumption of the Corollary implies that $BK_{1z} = K_{2z} B$. The assertion then follows from Theorem 1.6.4 if the A_j there are identified with the present K_{jz}.

In case A_1 and A_2 are similar self-adjoint operators, their unitary equivalence was noted by Toeplitz (cf. Wintner [3], p. 149). In case both operators are unitary the corresponding result is due to Wintner [3], p. 149. It was also pointed out by Wintner [4], pp. 139–140, that the unitary equivalence of normal operators is a consequence of their similarity equivalence when the underlying Hilbert space is finite-dimensional.

Remark. The polar factorization for bounded non-singular operators was first given by Wintner [3], p. 145. The result has a well-known generalization due to von Neumann [3], p. 307. See also Hartman [1], p. 233.

Wiegmann [1] showed that if A, B and AB are completely continuous normal operators on a Hilbert space then so also is BA. Kaplansky [1] has shown however that the normality alone of the bounded operators A, B and AB does not imply that of BA, although the implication does hold if, for instance, also either A or B is completely continuous. Kaplansky also showed, as a consequence of the result of Theorem 1.6.2, that if A and B are bounded and if A and AB are normal then B commutes with $A^* A$ if and only if BA is normal.

For some other applications of, or results related to, Fuglede's theorem, see also Beck and Putnam [1], Berberian [1, 4], Kaplansky [2], Kurepa [1, 2, 3], Putnam [12], Sakai [1], Stampfli [1] (p. 1457).

§ 1.7 Operator equation BX—$XA = Y$

Let B denote a Banach algebra of elements $A, B, X, \ldots$ and containing a unit I. Let T denote the operator $T(X)$ of B into itself defined by

$$T(X) = BX - XA . \tag{1.7.1}$$

The nature of the transformation T has been investigated by several authors. Some of the results are the following.

(i) If $B = B(\mathfrak{H})$ is the algebra of bounded operators on a Hilbert space $\mathfrak{H}$ and if $B + B^* \leqq b < a \leqq A + A^*$ then T^{-1} exists as a bounded operator with an integral representation

$$T^{-1}(X) = - \int_0^\infty e^{tB} X e^{-tA} dt \tag{1.7.2}$$

in terms of elements of $B(\mathfrak{H})$ (Heinz [1], p. 427). See the formulation in Rosenblum [1] and the reference given there to Cordes [1] for an extension of (i).

(ii) For an arbitrary Banach algebra B with identity I, if A and B have disjoint spectra then T^{-1} exists as a bounded operator (on B) with a contour integral representation

$$T^{-1}(X) = (1/2\pi i) \int (B - zI)^{-1} X (zI - A)^{-1} dz , \tag{1.7.3}$$

where the integral is taken over the boundary of an appropriate neighborhood of the spectrum of A (Rosenblum [1]). (Cf. the remarks in the beginning of § 1.4.)

Rosenblum's proof of (ii) depends on the operational calculus of Dunford and Taylor (see Taylor [1]). The assertion (ii) generalizes a result of Rutherford [1], where B is the algebra of $n \times n$ matrices.

For two sets S_1 and S_2 of the complex plane let $S_1 - S_2$ denote the set of complex numbers $z = z_1 - z_2$ with $z_k \in S_k$. If B is the Banach algebra of endomorphisms on a Banach space, Kleinecke has shown that the spectrum of T of (1.7.1), as an operator on B, is the set $\text{sp}(B) - \text{sp}(A)$; cf. Rosenblum [1], p. 265.

If B denotes the algebra $B(\mathfrak{H})$ of bounded operators on a Hilbert space $\mathfrak{H}$ and, if in (1.7.1), $A = B$, then $T(I) = 0$ and T is singular. (Cf. also the result of Kleinecke mentioned in the preceding paragraph.) In particular, no such formulas as (1.7.2) or (1.7.3) can exist for the solution, of the commutator equation

$$T(X) = Q, \quad T(X) = AX - XA .\tag{1.7.4}$$

In case A in (1.7.4) is normal however, an operational calculus for T has been constructed by Freeman [1] who deduces sufficient conditions in order that (1.7.4) have a solution X in $B(\mathfrak{H})$. The solution is represented as a singular integral operator similar to one obtained by Friedrichs [1, 2] (cf. §§ 6.11 ff.) in his investigations on perturbations of the self-adjoint multiplication operator $T : f(x) \to xf(x)$ on $L^2(a, b)$.

For a generalized formulation of the Friedrichs theory and applications to the perturbation theory of self-adjoint operators, see Schwartz [1, 2], also Rejto [1, 2].

Chapter II

Commutators and spectral theory

§ 2.1 Introduction

Let A and B be bounded operators on a Hilbert space and suppose that

$$C = AB - BA .\tag{2.1.1}$$

Then it is easily shown that $\|C\| \leq 2\|A\|\,\|B\|$ and, by consideration of simple finite-dimensional examples, that the last inequality can become an equality (with $C \neq 0$). Thus, in general, the factor 2 cannot be replaced by a smaller number. In case A and C are self-adjoint and C is semi-definite, the inequality can be refined however. In fact in this case, as will be shown below,

$$\|C\| \leq (2/\pi)\|A\|\,\|B\| ,\tag{2.1.2}$$

where the constant $2/\pi$ is optimal. In this chapter, some related inequalities will also be derived. In addition, an investigation of the spectrum of A when A is self-adjoint and $C \geq 0$ will be made.

§ 2.2 Spectral properties

The following was proved in Putnam [29].

Theorem 2.2.1. *Let A and B be bounded, A be self-adjoint and let the commutator C of (2.1.1) be self-adjoint and satisfy either $C \geq 0$ or $C \leq 0$. Then*

$$\|C\| \leq (1/2\pi)d_B \text{ meas sp}(A) ,\tag{2.2.1}$$

where $d_B = \max(\text{Im}(B)) - \min(\text{Im}(B))$ *and* $\text{Im}(B) = (1/2i)(B - B^*)$.

As before, if J is any self-adjoint operator, $\max J$ ($\min J$) will denote the maximum (minimum) point of $\text{sp}(J)$. The measure refers to ordinary one-dimensional Lebesgue measure. Since $\|B\| = \|B^*\|$ it is clear that $d_B \leq 2\|B\|$. Since also meas $\text{sp}(A) \leq 2\|A\|$ it is seen that (2.2.1) implies (2.1.2). The optimal nature of the factor $2/\pi$ will be shown later (§ 6.12).

An equivalent formulation of Theorem 2.2.1 is the following.

Theorem 2.2.1'. *Let* A, B, C *be bounded self-adjoint operators for which* $AB - BA = iC$ *with* $C \geq 0$ *or* $C \leq 0$. *Then*

$$\|C\| \leq (1/2\pi)[\max \operatorname{sp}(B) - \min \operatorname{sp}(B)] \operatorname{meas} \operatorname{sp}(A).$$

Closely related to Theorem 2.2.1 is the following one of Putnam [21] involving unitary equivalence of two self-adjoint operators, the difference of which is semi-definite.

Theorem 2.2.2. *Let* J *denote a bounded self-adjoint operator, let* U *be unitary, and suppose that*

$$UJU^* - J = D, \text{ where either } D \geq 0 \text{ or } D \leq 0. \tag{2.2.2}$$

Then

$$\|D\| \leq (1/2\pi) d \operatorname{meas}(\operatorname{sp}(U)), \tag{2.2.3}$$

where $d = \max J - \min J$.

Here the measure is one-dimensional Lebesgue measure on the boundary of the unit circle $|z| = 1$.

First Theorem 2.2.1 will be proved, granting Theorem 2.2.2, and then Theorem 2.2.2 will be proved.

Proof of Theorem 2.2.1 granting Theorem 2.2.2. On taking adjoints in (2.1.1) one sees that $AJ - JA = -iC$ where $J = \operatorname{Im}(B)$. Consequently $(A + iI)J - J(A + iI) = -iC$ and, on multiplying on the right and left by $(A + iI)^{-1}$,

$$J(A + iI)^{-1} - (A + iI)^{-1}J = -i(A + iI)^{-1}C(A + iI)^{-1}.$$

If U denotes the unitary operator

$$U = (A - iI)(A + iI)^{-1} \quad (= I - 2i(A + iI)^{-1}),$$

the Cayley transform of A (von Neumann [1], cf. also, e.g., Sz.-Nagy [1]), then one obtains $UJU^* - J = 2Q^*CQ$ where $Q = (A - iI)^{-1}$. Put $D = 2Q^*CQ$ and identify the present J and Q with those of Theorem 2.2.2. Clearly $D \geq 0$ or $D \leq 0$ according as $C \geq 0$ or $C \leq 0$. According to (2.2.3), $\operatorname{meas} \operatorname{sp}(U) \geq 2\pi \|D\|/d$, with $d = d_B$. However,

$$\|Q^*CQ\| = \sup_{\|x\|=1} (Q^*CQx, x) \geq \|C\| \inf_{\|x\|=1} \|Qx\|^2.$$

But if A has the spectral resolution $A = \int \lambda dE_\lambda$ then

$$Q = \int (\lambda - i)^{-1} dE_\lambda \quad \text{and hence} \quad \|Qx\|^2 \geq (1 + \|A\|^2)^{-1} \|x\|^2.$$

Hence

$$\operatorname{meas} \operatorname{sp}(U) \geq 4\pi \|C\|/d(1 + \|A\|^2).$$

On defining ϕ by $e^{i\phi} = (\lambda - i)(\lambda + i)^{-1}$ with $-\infty < \lambda < \infty$ and $0 < \phi < 2\pi$, it is seen that $d\phi = 2(1 + \lambda^2)^{-1} d\lambda$ and hence, if

$$S = \{\phi \in (0, 2\pi): e^{i\phi} \in \mathrm{sp}\,(U)\}\,,$$

then

$$\mathrm{meas\ sp}\,(U) = \mathrm{meas}\ S = 2 \int_{\mathrm{sp}(A)} (1+\lambda^2)^{-1}\,\mathrm{d}\lambda \le 2\ \mathrm{meas\ sp}\,(A).$$

For arbitrary $t > 0$, put $A_t = tA$. Then $tC = A_t B - BA_t$ and so, by the preceding results,

$$2\ \mathrm{meas\ sp}\,(A_t) \ge 4\pi t \|C\|/d_B(1+t^2 \|A\|^2).$$

Since $\mathrm{meas\ sp}(A_t) = t\ \mathrm{meas\ sp}\,(A)$, division of the last inequality by t followed by letting $t \to 0$ yields (2.2.1) and the proof of Theorem 2.2.1 is complete.

Proof of Theorem 2.2.2. There is no loss of generality in supposing that $D \ge 0$. Let U have the spectral resolution $U = \int_0^{2\pi} e^{i\lambda}\,\mathrm{d}E_\lambda$. Let $S = \{\lambda \in [0, 2\pi]: e^{i\lambda} \in \mathrm{sp}\,(U)\}$ (closed) and let S_c denote the complement of S, that is, $S_c = [0, 2\pi] - S$. Let $f(\lambda)$ denote a function defined on $[0, 2\pi]$, of class C^1 and such that $f(\lambda) \equiv 0$ on S. In particular,

$$f(\lambda) = \sum_{-\infty}^{\infty} c_k e^{ik\lambda} \quad \text{with} \quad \Sigma |c_k| < \infty \quad \text{and} \quad \int_0^{2\pi} f(\lambda)\,\mathrm{d}E_\lambda = 0.$$

It is clear that

$$\sum_{-m}^{n} c_k U^k = \int_0^{2\pi} \left(\sum_{-m}^{n} c_k e^{ik\lambda} \right) \mathrm{d}E_\lambda$$

converges in the uniform norm topology, as $m, n \to \infty$, to the zero operator and consequently $c_0 D^{\frac{1}{2}} + \Sigma' c_k D^{\frac{1}{2}} U^k = 0$, where the prime indicates that $k = 0$ is to be omitted from the summation.

If x is an arbitrary element of $\mathfrak{H}$ the Schwarz inequality implies that

$$\|c_0 D^{\frac{1}{2}} x\|^2 \le (\Sigma' |c_k|^2)(\Sigma' \|D^{\frac{1}{2}} U^k x\|^2)\,. \tag{2.2.4}$$

It is seen that (2.2.2) implies

$$\sum_{k=0}^{n} U^k D U^{*k} = U^{n+1} J U^{*n+1} - J \tag{2.2.5}$$

and

$$\sum_{k=1}^{n} U^{*k} D U^k = J - U^{*n} J U^n\,. \tag{2.2.6}$$

Addition of these two equations gives

$$\sum_{-n}^{n} U^{*k} D U^k = U^{n+1} J U^{*n+1} - U^{*n} J U^n\,.$$

Hence,

$$\sum_{-n}^{n} \| D^{\frac{1}{2}} U^k x \|^2 \leq d(x, x) \quad \text{for} \quad n = 1, 2, \ldots,$$

and thus

$$\Sigma' \| D^{\frac{1}{2}} U^k x \|^2 \leq d(x, x) - (Dx, x).$$

If one chooses $x = x_n$ to be unit vectors satisfying $Dx_n - \| D \| x_n \to 0$ as $n \to \infty$, it now follows from (2.2.4) and the Parseval relation

$$\Sigma |c_k|^2 = (1/2\pi) \int_0^{2\pi} |f(\lambda)|^2 \, d\lambda$$

that

$$|(1/2\pi) \int_0^{2\pi} f(\lambda) d\lambda|^2 \| D \| \leq$$

$$\left[(1/2\pi) \int_0^{2\pi} |f(\lambda)|^2 d\lambda - |(1/2\pi) \int_0^{2\pi} f(\lambda) d\lambda|^2 \right] (d - \| D \|). \quad (2.2.7)$$

Next choose $f(\lambda) = f_n(\lambda)$, where $\{f_n\}$ is a uniformly bounded sequence of C^1 functions tending almost everywhere to the characteristic function $k(\lambda)$ of the set S_c. A relation similar to (2.2.7) then holds with $f(\lambda)$ replaced by $k(\lambda)$. Since

$$\int_0^{2\pi} k(\lambda) d\lambda = \int_0^{2\pi} k^2(\lambda) d\lambda = \text{meas } S_c = 2\pi - \text{meas } S,$$

the desired relation (2.2.3) follows and the proof of Theorem 2.2.2 is complete.

A point λ is said to be in the essential spectrum (a term due to Wintner, cf. Hartman [2], p. 487) of a self-adjoint operator if it is either a limit point of the spectrum or is in the point spectrum with an infinite multiplicity.

The next result was proved by Putnam [17].

Theorem 2.2.3. *Under the same assumptions as in Theorem 2.2.2 there holds the inequality*

$$\| D \| \leq \delta, \quad (2.2.8)$$

where δ denotes the difference of the maximum and minimum points of the essential spectrum of J.

Since, as is clear from a trace argument, $D \geq 0$ or $D \leq 0$ can hold for finite-dimensional spaces only if $D = 0$ it can be supposed that $\mathfrak{H}$ is infinite-dimensional and hence any (bounded) self-adjoint operator has a non-empty essential spectrum. Again it is clear that $D \geq 0$ can be supposed.

Proof. Let λ_0 denote the maximum point in the essential spectrum of J and let the eigenvalues of J (if any) greater than λ_0 be given by

$\lambda_1 > \lambda_2 > \ldots$. If x_1 is any eigenvector of J belonging to λ_1 then, by (2.2.2),

$$0 \le (Dx_1, x_1) = (UJU^* x_1, x_1) - \lambda_1 (x_1, x_1) \le 0,$$

and so $0 = (\lambda_1 I - UJU^*)^{\frac{1}{2}} x_1$. Consequently, $(\lambda_1 I - UJU^*) x_1 = 0$ and so x_1 is an eigenvector of UJU^* belonging to λ_1. Since λ_1 belongs to the spectra of J and UJU^* with the same (finite) multiplicity it follows that the eigenvectors of J and UJU^* belonging to λ_1 are identical. Similarly, the eigenvectors of J and UJU^* for each λ_n are identical. In a similar manner it can be shown that if μ_0 is the minimum point in the essential spectrum of J, the eigenvectors of J and UJU^* belonging to each eigenvalue less than μ_0 are identical.

Next, if $x \in \mathfrak{H}$, it can be written as $x = z + w$, where z is the projection of x on the space spanned by the eigenvectors of J outside $\mu_0 \le \lambda \le \lambda_0$ and w is in the orthogonal complement of this space. Since $Dz = 0$ then

$$(Dx, x) = (Dw, w) = (UJU^* w, w) - (Jw, w) \le (\lambda_0 - \mu_0) \| w \|^2 \le (\lambda_0 - \mu_0) \| x \|^2$$

and (2.2.8) follows. The proof is now complete.

It follows from (2.2.3) that if (2.2.2) holds with $D \ne 0$ then meas sp $(U) > 0$. Although the measure of J need not be positive, relation (2.2.8) implies that J has at least two points in its essential spectrum. That J may have exactly two such points can be seen from the following example. Let $A = (a_{ij})$ and $B = (b_{ij})$, where $i, j, = 0, \pm 1, \pm 2, \ldots$, be doubly infinite diagonal matrices for which $a_{ii} = 1$ if $i = 0, 1, 2, \ldots$ and $a_{ij} = 0$ otherwise and $b_{ii} = 1$ if $i = 1, 2, \ldots$ and $b_{ij} = 0$ otherwise. Then the spectrum of both A and B consists of 0 and 1 each of infinite multiplicity. Hence $A = UBU^*$ for some unitary U, for instance, U can be the two-sided shift determined by $x = (\ldots, x_{-1}, x_0, x_1, \ldots) \to Ux = (\ldots, x_0, x_1, x_2, \ldots)$. Also $A - B = D = (d_{ij})$ with $d_{00} = 1$ and $d_{ij} = 0$ otherwise, so that $D \ge 0$ and $\| D \| = 1$.

It follows from (2.2.3) that sp (U) is the entire circle $|\lambda| = 1$.

If A is self-adjoint with the spectral resolution $A = \int \lambda dE_\lambda$ then the set $\mathfrak{H}_a = \mathfrak{H}_a(A)$ of elements x in $\mathfrak{H}$ for which $\| E_\lambda x \|^2$ is an absolutely continuous function of λ is a subspace of $\mathfrak{H}$ which reduces the operator A; see Halmos [3], p. 104, also Kato [2], p. 240, Kuroda [2], p. 436. The restriction of A to $\mathfrak{H}_a, A_a \equiv A/\mathfrak{H}_a$ is called the absolutely continuous part of A and, in case $\mathfrak{H}_a = \mathfrak{H}$, the operator $A = A_a$ is said to be absolutely continuous.

Similar notions can be defined for a unitary operator U with the spectral resolution $U = \int_0^{2\pi} e^{i\lambda} dE_\lambda$.

It is evident that the space $\mathfrak{H}_a$ is always the zero space if $\mathfrak{H}$ is finite-dimensional.

The next theorem is due to Putnam [29].

Theorem 2.2.4. *Let A and B satisfy the conditions of Theorem 2.2.1. Then*

$$\mathfrak{L} \subset \mathfrak{H}_a(A) \cap \mathfrak{H}_a(\mathrm{Im}(B)) , \qquad (2.2.9)$$

where $\mathfrak{L}$ denotes the smallest subspace of $\mathfrak{H}$ reducing both A and $\mathrm{Im}(B)$ and containing the range of C.

Proof. In view of the symmetric nature of the hypothesis on A and $\mathrm{Im}(B)$ (note that $A\,\mathrm{Im}(B) - \mathrm{Im}(B)\,A = -iC$) it is sufficient to show that $\mathfrak{L} \subset \mathfrak{H}_a(A)$.

To this end, it will first be shown by an argument given in Putnam [9] that

$$\mathfrak{R}_C \subset \mathfrak{H}_a(A) . \qquad (2.2.10)$$

Let $\{E_\lambda\}$ be the spectral family belonging to A and let $E(\beta)$ denote the projection operator corresponding to any Borel set β of the real line. If $\varDelta$ is an interval then clearly

$$E(\varDelta)CE(\varDelta) = \int_\varDelta (\lambda - \lambda_0)\mathrm{d}E_\lambda\, BE(\varDelta) - E(\varDelta)B \int_\varDelta (\lambda - \lambda_0)\mathrm{d}E_\lambda , \qquad (2.2.11)$$

where λ_0 is an arbitrary constant. If λ_0 is chosen to be the midpoint of $\varDelta$ then

$$\left\| \int_\varDelta (\lambda - \lambda_0)\mathrm{d}E_\lambda \right\| \leqq \tfrac{1}{2}|\varDelta| ,$$

and one obtains, on forming inner products,

$$\| C^{\frac{1}{2}} E(\varDelta)x \|^2 \leqq 2 \cdot \tfrac{1}{2}|\varDelta|\, \|B\|\, \|E(\varDelta)x\|^2 .$$

(Clearly it can be supposed here that $C \geqq 0$.)

If β denotes any Borel set and if $\{\varDelta_1, \varDelta_2, \ldots\}$ denotes any denumerable covering of β by pairwise disjoint intervals then

$$\| C^{\frac{1}{2}} E(\beta)x \| \leqq \sum_k \| C^{\frac{1}{2}} E(\varDelta_k)x \| \leqq \|B\|^{\frac{1}{2}} \sum_k |\varDelta_k|^{\frac{1}{2}} \|E(\varDelta_k)x\| . \qquad (2.2.12)$$

An application of the Schwarz inequality then implies that

$$\| C^{\frac{1}{2}} E(\beta)x \| \leqq \|B\|^{\frac{1}{2}} (\Sigma |\varDelta_k|)^{\frac{1}{2}} (\Sigma \|E(\varDelta_k)x\|^2)^{\frac{1}{2}}.$$

It then follows from (2.2.12) that

$$\| C^{\frac{1}{2}} E(\beta)x \| \leqq \|B\|^{\frac{1}{2}} (\mathrm{meas}\ \beta)^{\frac{1}{2}} \|x\| ,$$

hence

$$\| E(\beta)C^{\frac{1}{2}} \| = \| C^{\frac{1}{2}} E(\beta) \| \leqq \|B\|^{\frac{1}{2}} (\mathrm{meas}\ \beta)^{\frac{1}{2}} .$$

In particular, relation (2.2.10) follows.

Next, let t be any real number not in $\mathrm{sp}\,(J)$, where $J = \mathrm{Im}\,(B)$, and let $J_t = J - tI$. It is seen that $AJ_t - J_t A = AJ - JA = -iC$ and hence $AJ_t^{-1} - J_t^{-1}A = iJ_t^{-1}CJ_t^{-1}$. It now follows as above that $\mathfrak{H}_a(A) \supset \mathfrak{R}_{J_t^{-1}CJ_t^{-1}} = \mathfrak{R}_{J_t^{-1}C}$. Since $t \notin \mathrm{sp}\,(J)$, also $s \notin \mathrm{sp}\,(J)$ for $|s-t|$ sufficiently small. Since $\mathfrak{H}_a(A)$ is a subspace and since $d^n/ds^n(J_s^{-1}) = n!\,J_s^{-n-1}$ then $\mathfrak{H}_a(A) \supset \mathfrak{R}_{J_t^{-m}C}$ for $m = 1, 2 \ldots$, as well as for $m = 0$. Thus $\mathfrak{H}_a(A)$ contains the smallest subspace reducing J_t^{-1} and containing $\mathfrak{R}_C$. Clearly this space is the least space reducing J and containing $\mathfrak{R}_C$. But the absolute continuity of $\|E_\lambda x\|^2$ implies that of $\|E_\lambda A^n x\|^2$ for $n = 0, 1, 2, \ldots$, and so one has that

$$\mathfrak{R} \subset \mathfrak{H}_a(A), \tag{2.2.13}$$

where $\mathfrak{R}$ is the least subspace reducing A and containing $\mathfrak{M}_B$, and where $\mathfrak{M}_B$ is the least subspace reducing $J = \mathrm{Im}\,(B)$ and containing $\mathfrak{R}_C$.

Finally, it will be shown that $\mathfrak{L} = \mathfrak{R}$, and the assertion $\mathfrak{L} \subset \mathfrak{H}_a(A)$ then will follow from (2.2.13). Clearly $\mathfrak{R} \subset \mathfrak{L}$. In order to prove $\mathfrak{R} \supset \mathfrak{L}$, note that $\mathfrak{R}$ is the closure of the linear manifold of vectors $y = \Sigma\, A^m J^n C x_{mn}$ $(m, n \geq 0)$ where $x_{mn} \in \mathfrak{H}$. But $\mathfrak{L}$ is the closure of the linear manifold spanned by the vectors $z = PCx$ with $x \in \mathfrak{H}$ and P a product of a finite number of factors of non-negative powers of A and J. It follows from $AJ - JA = -iC$ that every such vector z is also a vector of type y and hence $\mathfrak{L} \subset \mathfrak{R}$. This completes the proof of Theorem 2.2.4.

For some results related to those of this section see also §§3.2, 3.12.

§ 2.3 Absolute continuity and measure of spectrum

It was shown in § 2.2 that if $UJU^* - J = D$ with $D \geq 0$ or $D \leq 0$ and $D \neq 0$, then $\mathrm{meas}\,\mathrm{sp}\,(U) > 0$ and, furthermore, that it is possible that the essential spectrum of J contains just two points, in which case $\mathrm{meas}\,\mathrm{sp}\,(J) = 0$. It turns out however that in this latter case necessarily $\mathrm{meas}\,\mathrm{sp}\,(U) = 2\pi$. In fact there will be proved the following result.

Theorem 2.3.1. Let J, U and D be defined as in Theorem 2.2.2, and suppose that $\mathrm{meas}\,\mathrm{sp}\,(U) < 2\pi$. Then both $\mathfrak{H}_a(U)$ and $\mathfrak{H}_a(J)$ contain the least subspace of $\mathfrak{H}$ reducing J and U and containing the range of D.

Proof. The hypothesis implies that there exists some $\lambda = e^{i\theta}$, with θ real, such that λ is not in $\mathrm{sp}\,(U)$. If $U_\theta = e^{-i\theta}U$ then $U_\theta J U_\theta^* - J = D$ and 1 is not in $\mathrm{sp}\,(U_\theta)$, so that U_θ is the Cayley transform of a bounded self-adjoint operator A. As in the proof of Theorem 2.2.1, let $Q = (A - iI)^{-1}$ and define C by $D = 2Q^*CQ$, that is, $C = \frac{1}{2}Q^{*-1}DQ^{-1}$. Then a reversal of the earlier argument leads to

$$AJ - JA = -iC. \tag{2.3.1}$$

Since $\mathfrak{H}_a(A) = \mathfrak{H}_a(U_\theta) = \mathfrak{H}_a(U)$, it follows from Theorem 2.2.4 that $\mathfrak{H}_a(U) \cap \mathfrak{H}_a(J)$ contains the least subspace $\mathfrak{M}$ of $\mathfrak{H}$ reducing A and J and containing $\mathfrak{R}_C$. But $\mathfrak{R}_C = \mathfrak{R}_{Q^{*-1}D} = \mathfrak{R}_{(A+iI)D}$, and hence $\mathfrak{M}$ is identical with the least subspace of $\mathfrak{H}$ reducing A and J and containing $\mathfrak{R}_D$. The proof of Theorem 2.3.1 is now complete.

It is seen that if $UJU^* - J = D$ with $D \geq 0$ and $D \neq 0$ and if sp(U) is not the entire circle $|\lambda| = 1$ then necessarily J has an absolutely continuous part and, in particular, meas sp$(J) > 0$. If also 0 is not in the point spectrum of D, then J is absolutely continuous, that is, J coincides with its absolutely continuous part.

Theorem 2.3.2. *Under the hypotheses of Theorem 2.2.2 (and whether or not the measure condition of Theorem 2.3.1 is satisfied), still $\mathfrak{H}_a(U)$ contains the least subspace of $\mathfrak{H}$ reducing U and containing the range of D.*

Proof. It is sufficient to show that $\mathfrak{H}_a(U)$, which obviously reduces U, contains the range of D. Since $UJU^* - J = U(JU^*) - (JU^*)U$, it follows from an argument like that in the first part of the proof of Theorem 2.2.4 that if $U = \int e^{i\lambda} dE_\lambda$, then $DE(Z) = 0$ where Z is any Borel set of measure 0. Thus also $E(Z)D = 0$ and the assertion follows. (See Putnam [17].)

Let $A = (a_{ij})$ and $B = (b_{ij})$ be doubly infinite matrices defined by $a_{ij} = \delta_{ij}\lambda_j$ and $b_{ij} = \delta_{ij}\lambda_{j-1}$ where $\{\lambda_n\}$, for $n = 0, \pm 1, \pm 2, \ldots$, is any sequence of real numbers satisfying $|\lambda_n| <$ const. and $\lambda_n < \lambda_{n+1}$ for all n. Then $A = UBU^*$ where U is, for instance, the shift operator (see § 2.2). It is clear that $A - B = D \geq 0$. Also, 0 is not in the point spectrum of D, and hence the range of D is dense in $\mathfrak{H}$. Consequently the absolute continuity of U follows from Theorem 2.3.2.

In case U is the shift operator there have thus been established the (known) results that sp$(U) = \{\lambda : |\lambda| = 1\}$ (see § 2.2) and that U is absolutely continuous. See also § 2.9.

Remark. Concerning absolute continuity of unitary operators, the concept of absolute continuity of contractions in terms of unitary dilations (cf. § 4.8) has been given by Schreiber [3, 4], Sz.-Nagy and Foiaş [2], see also Saitô and Yoshino [1]. The basic properties of unitary dilations were given by Halmos [2], Sz.-Nagy [3]; see also Schäffer [1]. Some additional references are Schreiber [1,2], Sz.-Nagy [5,6], Sz.-Nagy and Foiaş [1].

§ 2.4 Absolute continuity and numerical range

A complex number z will be said to belong to the interior of the convex set $W = W_Q$ belonging to a bounded operator Q (see § 1.5) if z is in W and if one of the following conditions holds: if W is two-dimensional, z is an interior point; if W is a line segment, z is not an end-point; finally, W consists of the single point z.

It follows from Theorem 1.5.1 that if, for instance, A is normal or even semi-normal, then 0 is in W_C where $C = AB - BA$. In general however, 0 need not belong to W_C (Halmos [4]); see § 1.5 above. The following also concerns the set W_C and was proved in Putnam [9].

Theorem 2.4.1. (i) *Suppose that* $AB-BA=C$, *where A and B are bounded, and that A is normal with the spectral resolution*

$$A = \int z \, dK_z . \tag{2.4.1}$$

Suppose that there exists a Borel set S in the complex plane with the property that

$$K(S) = I , \tag{2.4.2}$$

and that for every $\varepsilon > 0$ there exists a sequence of pairwise disjoint Borel sets $\beta_1, \beta_2, \ldots$ such that $S \subset \cup \beta_n$ and $\Sigma \operatorname{diam} \beta_n < \varepsilon$. Then 0 (which lies in W_C) belongs to the interior of the set $W = W_C$. (ii) *If A is self-adjoint or unitary and if there exists a Borel set S of one-dimensional measure zero for which (2.4.2) holds then 0 belongs to the interior of W_C.*

Proof of (i). The proof is similar to one given in § 2.2. Choose θ so that the set $W_\theta = W_{C_\theta}$, where $C_\theta = e^{i\theta} C = A(e^{i\theta} B) - (e^{i\theta} B)A$, lies in the half-plane $\operatorname{Re}(z) \geqq 0$. Then, corresponding to (2.2.11), one has

$$K(\beta) C_\theta K(\beta) = \int_\beta (z - z_0) dK_z B_\theta K(\beta) - K(\beta) B_\theta \int_\beta (z - z_0) dK_z, \tag{2.4.3}$$

where $B_\theta = e^{i\theta} B$, β is any Borel set, and z_0 is an arbitrary complex number. On choosing z_0 in β, taking adjoints, and putting $H = C_\theta + C_\theta^*$, one obtains (cf. (2.2.12))

$$\| H^{\frac{1}{2}} K(\beta) x \| \leqq \operatorname{const.} (\operatorname{diam} \beta)^{\frac{1}{2}} \| K(\beta) x \| . \tag{2.4.4}$$

If now the sets β_n are defined as in the theorem then, using (2.4.2), $H^{\frac{1}{2}} = H^{\frac{1}{2}} K(S) = H^{\frac{1}{2}} \Sigma K(S\beta_n)$, and one obtains by virtue of the Schwarz inequality,

$$\| H^{\frac{1}{2}} x \| \leqq \operatorname{const.} (\Sigma \operatorname{diam}(S\beta_n))^{\frac{1}{2}} (\Sigma \| E(S\beta_n) x \|^2)^{\frac{1}{2}} . \tag{2.4.5}$$

But $\Sigma \| E(S\beta_n) x \|^2 = \| x \|^2$ and so (2.4.5) implies that $\| H^{\frac{1}{2}} \| \leqq \operatorname{const.} \varepsilon^{\frac{1}{2}}$ for every $\varepsilon > 0$, that is, $H = 0$. But this means that

$$Ce^{i\theta} = -(Ce^{i\theta})^* \tag{2.4.6}$$

so that $Ce^{i(\theta + \pi/2)}$ is self-adjoint. Since $W_{\theta + \pi/2}$ is a segment of the real axis, the set $W(=W_C)$ contains the origin. If 0 is not in the interior of this segment there exists an angle ϕ for which the set W_ϕ belonging to $Ce^{i\phi}$ lies on the real axis with left end-point at 0. In view of (2.4.6), which now holds for $\theta = \phi$, together with the fact that $Ce^{i\phi}$ is self-adjoint, it follows that $Ce^{i\phi} = 0$ and hence $C = 0$. But then W_C is the single point 0 and 0 is, by definition, in the interior of W_C, a contradiction. This completes the proof of (i).

Proof of (ii). Note that (2.4.4) holds for any Borel set β. It is then easily proved from this result (cf. also (2.4.5)) that if the set S is a subset either of a segment or of the boundary of a circle then

$$\| H^{\frac{1}{2}} K(S) \| \leq \text{const. (meas } S)^{\frac{1}{2}} , \tag{2.4.7}$$

where the measure refers to the ordinary one-dimensional measure of S. Since meas $S=0$, the remainder of the proof of (ii) is similar to that of (i) and can be omitted.

As a consequence of the proof of (ii) of Theorem 2.4.1 there holds the following result.

Theorem 2.4.2. *Let A be bounded and self-adjoint or unitary, B bounded and arbitrary, and $C=AB-BA$. Suppose that 0 is not in the numerical range of C (although, by Theorem 1.5.1, it is in its closure W_C), so that in particular 0 is on the boundary of W_C. Then A is absolutely continuous.*

Proof. If S is any Borel set of measure 0, it is seen that $H^{\frac{1}{2}} K(S)=0$. Hence also $HK(S)=0$ and $K(S)H=0$. But the assumptions imply that 0 is not in the point spectrum of H, so that $\mathfrak{R}_H$ is dense, and hence $K(S)=0$ as was to be shown.

For some related results, see Putnam [15].

§ 2.5 Higher order commutators

For A and B bounded, the n-th order commutator $B^{(n)}$ is defined by $B^{(n)}=AB^{(n-1)}-B^{(n-1)}A$ where $B^{(0)}=B$ and $B^{(1)}=AB-BA$. It was noted earlier that commutators act like differentiations; see § 1.3. Let C, D, E denote the first, second and third commutators respectively, so that

$$C=AB-BA, \quad D=AC-CA, \quad E=AD-DA . \tag{2.5.1}$$

The following theorems were given in Putnam [13].

Theorem 2.5.1. *Let the bounded operators A, B, C, D satisfy (2.5.1) and let A be normal with the spectral resolution (2.4.1). Suppose that there exists a Borel set S of two-dimensional zero measure with the property that $K(S)=I$. Then 0 is in the interior of the set W_D.*

Proof. The proof is similar to that of Theorem 2.4.1. Choose θ so that the set $W_{D e^{i\theta}}$ belonging to D_θ, where

$$D_\theta = AC_\theta - C_\theta A \qquad (C_\theta = Ce^{i\theta}, \ D_\theta = De^{i\theta}) , \tag{2.5.2}$$

lies in the half-plane $\text{Re}(z) \geq 0$. Then $J \equiv D_\theta + D_\theta^* \geq 0$. Multiplications on the right and left sides of (2.5.2) by $K(\beta)$, where β is any Borel set, yield

$$K(\beta) D_\theta K(\beta) = \int_\beta (z-z_0) dK_z \, C_\theta K(\beta) - K(\beta) C_\theta \int_\beta (z-z_0) dK_z .$$

It follows (cf. (2.4.3)) that $\|J^{\frac{1}{2}}K(\beta)x\|^2 \leq \text{const.}\,(\text{diam }\beta)^2\,\|K(\beta)x\|^2$ and hence, arguing as before, $\|J^{\frac{1}{2}}K(S)\| \leq \text{const.}\,(\Sigma(\text{diam }\beta_n)^2)^{\frac{1}{2}}$, where $\{\beta_n\}$ is a sequence of disjoint Borel sets covering S. But, since S has two-dimensional measure 0, the right side of the last inequality can be made arbitrarily small and hence, since $K(S)=I$, $J=0$. It follows as in the proof of Theorem 2.4.1 that 0 is in the interior of W_D, a contradiction, and the proof is complete.

Theorem 2.5.2. *Let the bounded operators A, B, C, D satisfy (2.5.1) and let A be normal with the spectral resolution (2.4.1). Suppose that 0 is not in the numerical range of D, so that in particular 0 is on the boundary of W_D. Then A is absolutely continuous in the sense of two-dimensional measure, that is,*

$$K(S) = 0 \tag{2.5.3}$$

for any Borel set S of two-dimensional measure 0.

Proof. As in the preceding proof one has $K(S)J=0$. Since 0 is not in the point spectrum of J then $\Re_J$ is dense and (2.5.3) follows.

Finally, in this section, there will be proved the following.

Theorem 2.5.3. *If A is normal then 0 always belongs to the interior of the set W_E, where E is defined by (2.5.1).*

Proof. An argument similar to that used in proving Theorem 2.4.1 shows that if the theorem is not true, then θ can be chosen so that $L_\theta = E_\theta + E_\theta^* \geq 0$, where $E_\theta = Ee^{i\theta}$. Then for $S = \text{sp}(A)$,

$$\|L_\theta^{\frac{1}{2}}K(S)x\| \leq \text{const.}\,\|x\|\,(\Sigma(\text{diam }\beta_n)^3)^{\frac{1}{2}}, \tag{2.5.4}$$

where the β_n are defined as before. Since the summation can be made arbitrarily small and since $K(S)=I$, then $L_\theta=0$ and, as before, a contradiction is obtained.

§ 2.6 Further results on commutators and normal operators

It was shown above that if $C=AB-BA$ and if A is self-adjoint with the spectral resolution $A = \int \lambda\,dE_\lambda$ and if $E(\beta)=I$ for some Borel set β of (one-dimensional) measure 0, then necessarily 0 is in the interior of the set W_C. Since there exist examples of the form

$$C = AA^* - A^*A = (A+A^*)A^* - A^*(A+A^*)$$

where $C \geq 0$ and $C \neq 0$ and where, necessarily, the spectrum of the self-adjoint operator $A+A^*$ has a (one-dimensional) positive measure (see Theorem 2.2.1) it is clear that the last assertion can become false if "self-adjoint" and "one-dimensional" are replaced by "normal" and "two-dimensional." However the assertion can be made for certain types of normal operators with spectra of two-dimensional zero measure. The following discussion is based on Putnam [14].

Let T be a bounded self-adjoint operator and let $\{\Delta_k\}$ denote any covering of $\mathrm{sp}\,(T)$ by a finite number of pairwise disjoint intervals Δ_k of length d_k. For each $\delta > 0$ define the positive number $f(T, \delta)$ by

$$f(T, \delta) = \inf \left(\sum_k d_k \right), \qquad d_k \geq \delta . \tag{2.6.1}$$

Theorem 2.6.1. *Let A and B be bounded, A normal and B arbitrary, and let $C = AB - BA$. Let A be represented in its Cartesian form*

$$A = H + iJ, \quad H = \tfrac{1}{2}(A + A^*), \quad J = (1/2i)(A - A^*) . \tag{2.6.2}$$

Then

$$0 \text{ is in the interior of } W_C \tag{2.6.3}$$

whenever

$$\inf_{\delta, \eta > 0} \; [f(H, \delta) f(J, \eta)(\delta^{-1} + \eta^{-1})] = 0 . \tag{2.6.4}$$

Proof. Assume (2.6.4). If (2.6.3) is false, there exists some real θ such that $M = \mathrm{Re}\,(Ce^{i\theta}) \geq 0$ and $M \neq 0$. It will be shown that $M = 0$, a contradiction.

It was shown in § 2.4 (cf. (2.4.5)) that if $\{\gamma_k\}$ is any covering of the spectrum of the normal operator A by pairwise disjoint Borel sets then

$$\|M\| \leq \text{const.} \; \Sigma \,(\mathrm{diam}\; \gamma_k) . \tag{2.6.5}$$

(Note that $\|M\| = \|M^{\frac{1}{2}}\|^2$.) Let $\delta, \eta > 0$ and let $\{\Delta_k\}$ and $\{\varepsilon_k\}$ denote finite coverings, each consisting of pairwise disjoint intervals, of $\mathrm{sp}\,(H)$ and $\mathrm{sp}\,(J)$ respectively, and such that $d_k = |\Delta_k| \geq \delta$ and $e_k = |\varepsilon_k| \geq \eta$. Since A is normal, $\mathrm{sp}\,(A)$ is contained in the product set $\mathrm{sp}\,(H) \times \mathrm{sp}\,(J)$ and hence, by (2.6.5),

$$\|M\| \leq \text{const.} \; \sum_j \sum_k (d_j^2 + e_k^2)^{\frac{1}{2}} .$$

But this double summation is majorized by

$$\sum_j \sum_k (d_j + e_k) \leq \sum_j d_j \sum_k e_k (\delta^{-1} + \eta^{-1}),$$

and so $\|M\|$ is not greater than a constant times the expression on the left side of (2.6.4). Hence if (2.6.4) holds then $M = 0$ and the proof is complete.

An application of the result will next be given. Let $0 < p < 1$ and let C_p denote the Cantor set obtained by first removing from $0 \leq \lambda \leq 1$ the open middle portion of length p and then successively removing the open middle p-th part of the remaining intervals. For $p = 1/3$ one obtains the standard Cantor set.

Corollary. *Let A, B, C be defined as in Theorem 2.6.1 and let $\mathrm{sp}\,(H) = C_p$ and $\mathrm{sp}\,(J) = C_q$ where $\tfrac{1}{2} < p, q < 1$. Then (2.6.3) holds.*

Proof. At the n-th stage of construction of C_p the length of each of the 2^n unremoved intervals is $2^{-n}(1-p)^n$. The union of these intervals forms a covering $\{\varDelta_k\}$ of $C_p = \text{sp}(H)$ by pairwise disjoint intervals, of total length $(1-p)^n$, with $\delta = 2^{-n}(1-p)^n$. A similar situation exists for the m-th stage of construction of C_q and so the bracketed expression of (2.6.4) is not greater than

$$(1-p)^n(1-q)^m\left[2^n(1-p)^{-n}+2^m(1-q)^{-m}\right] = 2^n(1-q)^m+2^m(1-p)^n.$$

If $m=n\to\infty$ then, by virtue of the hypothesis $\frac{1}{2}<p,q<1$, it is seen that (2.6.4) holds, and the Corollary follows.

It is clear that (2.6.4) implies in particular that

$$\text{meas sp}(H) = 0 \quad \text{and} \quad \text{meas sp}(J) = 0. \tag{2.6.6}$$

Whether (2.6.6) implies (2.6.3) will remain undecided. However the following assertion can be made.

Theorem 2.6.2. *Let $\mathfrak{H}$ be separable and let A, B, C be defined as in Theorem 2.6.1. If meas sp$(H) = 0$ and if J has a pure point spectrum then (2.6.3) holds.*

Proof. If $A = \int z\,dK_z$ is the spectral resolution of A then $K(S) = I$ where S is the product set of sp(H) and, since $\mathfrak{H}$ is separable, an at most denumerable set. As in the proof of Theorem 2.6.1 it is clearly enough to show the existence of coverings $\{\gamma_k\}$ of S by pairwise disjoint Borel sets for which the right side of (2.6.5) can be made arbitrarily small. That such coverings exist is clear and the proof is complete.

§ 2.7 Half-bounded operators and unitary equivalence

Most of the material presented so far has dealt with bounded operators. The following perturbation theorem was proved by Putnam [21] and deals with half-bounded self-adjoint operators.

Theorem 2.7.1. *Let J and D denote non-negative self-adjoint operators on a Hilbert space $\mathfrak{H}$ and suppose that D is bounded. Let $B = J+D$ and suppose that $J+D$ and J are unitarily equivalent, so that $J+D = UJU^*$ with U unitary. Let x be any element of $\mathfrak{H}$ for which $y = D^{\frac{1}{2}}x \neq 0$ and $y \in \mathfrak{D}_J$. Then*

$$\text{meas sp}(U) \geq 2\pi\left[1+2\|x\|^2(Jy,y)/\|y\|^4\right]^{-1}. \tag{2.7.1}$$

Proof. The proof will depend upon a modification of the argument used in proving Theorem 2.2.2. Let $f(\lambda)$ be real, of period 2π, and of class C^1, and put

$$f(\lambda) = c_0 + g(\lambda) + \bar{g}(\lambda), \quad g(\lambda) = \sum_{k=1}^{\infty} c_k e^{ik\lambda}. \tag{2.7.2}$$

Let x and y be defined as in the statement of the theorem. Then

$$\left(\int_0^{2\pi} g(\lambda) dE_\lambda y, D^{\frac{1}{2}} x \right) = \left(\sum_1^\infty c_k D^{\frac{1}{2}} U^k y, x \right). \qquad (2.7.3)$$

On taking conjugates and using (2.7.2), it is seen that

$$\int_0^{2\pi} f(\lambda) d \| E_\lambda y \|^2 = c_0 \| y \|^2 + 2 \operatorname{Re} \left(\sum_1^\infty c_k D^{\frac{1}{2}} U^k y, x \right). \qquad (2.7.4)$$

If now $f(\lambda)$ is chosen as in § 2.2 so as to be 0 on the set

$$S = \{ \lambda \in [0, 2\pi] : e^{i\lambda} \in \operatorname{sp}(U) \},$$

then the left side of (2.7.4) is 0 and hence

$$|c_0|^2 \| y \|^4 \leq 4 \| x \|^2 \sum_1^\infty |c_k|^2 \sum_1^\infty \| D^{\frac{1}{2}} U^k y \|^2. \qquad (2.7.5)$$

Since D is bounded, relation (2.2.6) is a valid operator equation (applicable on $\mathfrak{D}_J$). Since $J \geq 0$ and $y \in \mathfrak{D}_J$ it follows from (2.2.6) that

$$\sum_1^\infty \| D^{\frac{1}{2}} U^k y \|^2 \leq (Jy, y). \qquad (2.7.6)$$

Hence, by (2.7.5),

$$|c_0|^2 \| y \|^4 \leq 4 \| x \|^2 \sum_1^\infty |c_k|^2 (Jy, y). \qquad (2.7.7)$$

The remainder of the argument is similar to that given in the proof of Theorem 2.2.2 and will be omitted.

Applications of Theorem 2.7.1 to wave operators will be given later (Chapter V).

§ 2.8 Half-boundedness and absolute continuity

Theorem 2.8.1. *Let J, D and B be defined as in Theorem 2.7.1. Then $\mathfrak{H}_a(U)$ contains the least subspace of $\mathfrak{H}$ reducing U and containing the range of D.*

Proof. Let x be arbitrary, $y = D^{\frac{1}{2}} x$, and z belong to $\mathfrak{D}_J$. Then if

$$h(\lambda) = \sum_{k=0}^\infty c_k e^{ik\lambda}$$

is of class C^1 one has (cf. (2.7.3))

$$\int_0^{2\pi} h(\lambda) d(E_\lambda z, y) = c_0(z, y) + \left(\sum_{k=1}^{\prime\infty} c_k D^{\frac{1}{2}} U^k z, x \right), \qquad (2.8.1)$$

and hence, by an argument similar to that used above,

$$\left| \int_0^{2\pi} h(\lambda) d(E_\lambda z, y) \right| \leqq |c_0| \|z\| \|y\| + \|x\| \left(\sum_{k=1}^{\infty} |c_k|^2 \right)^{\frac{1}{2}} (Jz, z)^{\frac{1}{2}} . \quad (2.8.2)$$

Consequently, using the Schwarz inequality and the Parseval relation,

$$\left| \int_0^{2\pi} h(\lambda) d\sigma(\lambda) \right|^2 \leqq C_{xz} \int_0^{2\pi} |h(\lambda)|^2 d\lambda , \quad \sigma(\lambda) = (E_\lambda z, y) , \quad (2.8.3)$$

where C_{xz} denotes a number depending on x and z but not on $h(\lambda)$.

If $L(h) = \int_0^{2\pi} h(\lambda) d\sigma(\lambda)$, with h as above, it is clear that $L(h)$ is a linear functional defined on a dense set of the subspace $\mathfrak{H}^+$ of $L^2(0, 2\pi)$ where

$$\mathfrak{H}^+ = \left\{ x = \sum_{k=0}^{\infty} a_k e^{ik\lambda} : \Sigma |a_k|^2 < \infty \right\} .$$

It follows from (2.8.3) that for h of class $C^1, |L(h)| \leqq \text{const.} \|h\|$. If now h is an arbitrary element of $\mathfrak{H}^+$ choose $h_n \in C^1$ (h_n in $\mathfrak{H}^+$) so that $\|h_n - h\| \to 0$. Since $|L(h_n) - L(h_m)| \leqq \text{const.} \|h_n - h_m\|$, it follows that $L(h) \equiv \lim_{n \to \infty} L(h_n)$ exists. Moreover it is clear that this definition of $L(h)$ is independent of the particular sequence $\{h_n\}$ chosen to approximate h and that $L(h)$ is a bounded linear functional on $\mathfrak{H}^+$. Hence, by the Fréchet-Riesz theorem (cf. Riesz and Sz.-Nagy [1], p. 61), $L(h) = (h, k)$ for some k in $\mathfrak{H}^+$. Hence,

$$\int_0^{2\pi} h(\lambda) d\left[\sigma(\lambda) - \int_0^\lambda \bar{k}(\mu) d\mu \right] = 0 \quad (2.8.4)$$

for all h in $\mathfrak{H}^+$ and of class C^1 and, in particular, for $h = e^{in\lambda}$ ($n = 0, 1, 2, \ldots$). Hence, by a form of the F. and M. Riesz theorem (F. and M. Riesz [1], although the result is there formulated differently; see also Halmos [1], Sarason [1]), the function $\sigma(\lambda) - \int_0^\lambda \bar{k}(\mu) d\mu$, and hence also $\sigma(\lambda)$, is absolutely continuous.

Thus $(E_\lambda z, y)$ is absolutely continuous whenever $z \in \mathfrak{D}_J$ and $y \in \mathfrak{R}_{D^{\frac{1}{2}}}$. Since $\mathfrak{D}_J$ is dense, it follows that $\|E_\lambda x\|^2$ is absolutely continuous for all $x \in \mathfrak{R}_{D^{\frac{1}{2}}}$. But $[\mathfrak{R}_{D^{\frac{1}{2}}}] = [\mathfrak{R}_D]$, where $[\mathfrak{M}]$ denotes the closure of the linear manifold $\mathfrak{M}$, and hence $\|E_\lambda x\|^2$ is absolutely continuous for all x in $\mathfrak{R}_D$. Clearly then $\|E_\lambda x\|^2$ is absolutely continuous for all x in the closure of the linear manifold of finite linear combinations of elements x_n in $\mathfrak{R}_{U^n D}$ ($n = 0, \pm 1, \pm 2, \ldots$) and the proof is complete.

Remark. In order to reduce (2.8.4) to an explicit form of a theorem given in F. and M. Riesz [1], note that it is possible to choose a constant c so that

$$\eta(\lambda) = \sigma(\lambda) - \int_0^\lambda \bar{k}(\mu) d\mu + c\lambda$$

satisfies $\eta(0)=\eta(2\pi)$. Clearly $\eta(\lambda)$ is of bounded variation and (2.8.4) implies

$$\int_0^{2\pi} e^{in\lambda}\,d\eta(\lambda) = 0 \qquad (n=1, 2, \ldots) , \tag{2.8.5}$$

and hence, on integrating by parts,

$$\int_0^{2\pi} \eta(\lambda)\,e^{in\lambda}\,d\lambda = 0 \qquad (n=1, 2, \ldots). \tag{2.8.6}$$

But this implies that $\eta(\lambda)\in\mathfrak{H}^+$ and the absolute continuity of $\eta(\lambda)$ then follows from F. and M. Riesz [1], pp. 33 ff.

The assertion of Theorem 2.8.1 was stated in Putnam [30], see also Putnam [23]. However, the $F(\lambda)$ in formula line (24) of [30] should be replaced by $F^+(\lambda)$. The proof can then be completed as above.

An immediate consequence of Theorem 2.8.1 is the following.

Corollary. *Let J, D and B be defined as in Theorem 2.7.1 and suppose that 0 is not in the point spectrum of D. Then U is absolutely continuous.*

§ 2.9 Applications

There was established in §§ 2.2 and 2.3 above, using results on commutators, the known fact that the shift operator $(Ux)_n = x_{n+1}$ on the doubly infinite sequential Hilbert space satisfies

$$U \text{ is absolutely continuous and } \operatorname{sp}(U) = \{\lambda:|\lambda| = 1\} . \tag{2.9.1}$$

Relation (2.9.1) will be proved for other unitary operators U in the next several examples. See Putnam [21, 24].

(i) Consider first the one-dimensional space $L^2(-\infty, \infty)$ of functions $x=x(t)$. Corresponding to the shift operator mentioned above is the translation operator U on L^2 defined by

$$U:x(t) \to x(t+\alpha), \quad -\infty < t < \infty , \tag{2.9.2}$$

where α is a constant, which will be supposed positive. It is obvious that U is unitary. Moreover it is well-known that (2.9.1) holds. (Actually U can be expressed as $U=e^{i\alpha p}$ where p is the momentum operator of quantum mechanics; cf. § 4.2.) A proof of (2.9.1) using methods of this chapter will be given below.

For $\beta=$const. >0, let B and J denote the bounded multiplication operators defined by

$$B = \operatorname{Arc\,tan}(\beta^{-1}(t+\alpha))+\pi/2, \quad J = \operatorname{Arc\,tan}(\beta^{-1}t)+\pi/2 . \tag{2.9.3}$$

It is readily verified that B and J are non-negative operators, that $B=UJU^*$, and that $B-J=D$ is a multiplication operator $d(t)$ satisfying $d(t)>0$. Since $D\geq 0$ and 0 is not in the point spectrum of D it

follows from Theorem 2.3.2 that U is absolutely continuous, so that
the first part of (2.9.1) is proved. The second assertion will follow from
Theorem 2.7.1 if it is shown that

$$\inf_{y}\left\{\int_{-\infty}^{\infty} d^{-1}y^2\,dt \int_{-\infty}^{\infty}(\text{Arc tan}\,(\beta^{-1}t)+\pi/2)y^2\,dt \bigg/ \left(\int_{-\infty}^{\infty}y^2\,dt\right)^2\right\} = 0 \tag{2.9.4}$$

where y, $d^{-\frac12}y$ are real and belong to $L^2(-\infty, \infty)$.

To this end, choose δ so that $0<\delta<\frac14\alpha$ and then y so that $y\equiv 0$
outside $[-2\delta, -\delta]$ and $0<\int_{-\infty}^{\infty}y^2\,dt<\infty$. Then $t+\alpha\geq\frac12\alpha$ for $-2\delta\leq$
$t\leq -\delta$. Clearly for any $\varepsilon>0$ there exists a constant $\beta>0$ so small
that $0\leq\text{Arc tan}\,(\beta^{-1}t)+\pi/2<\varepsilon$ and $d(t)>\pi-\varepsilon$ for $-2\delta\leq t\leq -\delta$.
Hence there exist y for which $\{\ldots\}$ of (2.9.4) can be made arbitrarily
small and so (2.9.4) holds.

Thus U of (2.9.2) satisfies (2.9.1).

(ii) A generalization of the situation of (i) is the following. Consider
the conservative, vector system of differential equations

$$x' = f(x), \quad f \text{ of class } C^1_* \text{ on } \Omega, \tag{2.9.5}$$

where Ω is some connected open set of Euclidean n-space. In addition
suppose that Ω is an invariant set of the system, so that if $x_0\in\Omega$ then the
solution $x=x(t)$ of (2.9.5) exists and lies in Ω for $-\infty<t<\infty$. Finally
suppose that (2.9.5) is incompressible, so that

$$\text{div}\,f = 0. \tag{2.9.6}$$

(Condition (2.9.6) assures the invariance of Lebesgue measure; for a
succinct discussion see Cesari [1], pp. 103–104.) Then $T_t:x=x(0)\rightarrow x(t)$
determines on $L^2(\Omega)$ a unitary transformation (cf., e.g., Hopf [1]) $U=U_t$
defined by

$$U_t g(x) = g(T_t x). \tag{2.9.7}$$

Let $E=\{x\in\Omega : f(x)=0\}$, the set of equilibrium points of (2.9.5). Clearly
$L^2(E)$ reduces U and the restriction of U to E is the identity. It is clear
that $E\neq\Omega$ if and only if meas $(\Omega-E)>0$.

Theorem 2.9.1. *Under the above hypotheses suppose that there
exists a function* $\phi=\phi(x)$ *of class* C^2 *on* Ω *for which*

$$f = \text{grad}\,\phi. \tag{2.9.8}$$

Then (1) *either* $\Omega=E$ *and hence* $U_t=I$ *for all* t *or* (2) E *is a set of measure
zero and* $U=U_t$ *satisfies* (2.9.1) *for* $t\neq0$.

Proof. Suppose that $E\neq\Omega$ and that $t>0$. Then by (2.9.5) and (2.9.8),
$d\phi/dt=|\text{grad}\,\phi|^2$ along any solution path of (2.9.5). Hence

$$\phi(T_t x)-\phi(x) = \int_0^t |\text{grad}\,\phi|^2\,du \quad (x=x_0). \tag{2.9.9}$$

Let $\beta > 0$ and let B, J denote the non-negative bounded multiplication operators

$$B = \text{Arc} \tan (\beta^{-1} \phi(T_t x)) + \pi/2, \quad J = \text{Arc} \tan (\beta^{-1} \phi(x)) + \pi/2 . \tag{2.9.10}$$

Then clearly $B = UJU^*$ and $B - J = D$ where D is a multiplication operator $d(x, t)$. It follows from (2.9.9) and (2.9.10) that $d(x, t) > 0$ for $t > 0$ and $x \in \Omega - E$. Hence by Theorem 2.3.2 the restriction of $U = U_t$ to $\Omega - E$ is for $t > 0$ (and, since $U_{-t} = U_t^{-1}$, also for $t < 0$) absolutely continuous. An argument similar to that used above (cf. (2.9.4)) shows that also meas $\text{sp}(U_t) = 2\pi$ for $t > 0$ (hence for $t < 0$) and so (2.9.1) holds for U_t $(t \neq 0)$ defined by (2.9.7).

It follows from (2.9.6) and (2.9.8) that ϕ is harmonic in Ω. If ψ denotes any one of the components of grad ϕ then ψ is also harmonic in Ω and $\psi = 0$ on E. Since ψ is a real analytic function of n variables then either $\psi \equiv 0$ in Ω or meas $\{x : \psi(x) = 0\} = 0$, where "meas" here refers to Lebesgue volume measure on Ω; cf. the footnote in Putnam [24], p. 389. Hence if meas $E > 0$ it follows that $\psi \equiv 0$ in Ω and so $\Omega = E$, a contradiction. Hence meas $E = 0$ and the proof of the theorem is complete.

In case $n = 1$, relation (2.9.6) implies that $f \equiv a$ ($= \text{const.}$) and so (2.9.8) holds with $\phi = ax$. Since, in this case, the solutions of (2.9.5) are $x(t) = at + x_0$ this example is essentially that treated in (i) above.

In case $n = 2$, the condition (2.9.6) implies that the system (2.9.5) is Hamiltonian (cf. Wintner [7], p. 88), thus there exists a function $H = H(x_1, x_2)$ such that $x_1' = \partial H/\partial x_2$, $x_2' = -\partial H/\partial x_1$. Condition (2.9.8) is then fulfilled in case H is also harmonic.

If (2.9.6) and (2.9.8) hold, and if $\Omega \neq E$, it was noted that $\Omega - E = \Omega$ and that $U = U_t$ $(t \neq 0)$ is absolutely continuous on Ω. In particular U has no point spectrum and so meas $\Omega = \infty$. In fact, if $0 < \text{meas } \Omega < \infty$, the characteristic function of Ω would be an eigenfunction of U belonging to the eigenvalue 1. It can be noted that when Ω is simply connected the condition (2.9.8) holds if the flow is irrotational. Compare for $n = 3$ the assertion of Theorem 2.9.1 with a problem in Kellogg [1], p. 215, Ex.2.

(iii) Let T denote a μ-measure preserving transformation on a space Ω of points x. Then the transformation

$$U : f(x) \rightarrow f(Tx) \tag{2.9.11}$$

is unitary on the space $L^2(\Omega, d\mu)$. The transformation T is said to be dissipative if there exists a set A of positive measure for which the images $A_n = T^n(A)$ $(n = 0, \pm 1, \pm 2, \ldots)$ are disjoint and $\Omega = \bigcup_{-\infty}^{\infty} A_n$; cf. Hopf [1], p. 46, Halmos [6], p. 11. Such a set A will be called a generating set of Ω.

Theorem 2.9.2. *If T is dissipative on Ω then U of (2.9.11) satisfies (2.9.1).*

Proof. As in § 2.3, let $\{\lambda_n\}$, $n=0$, $\pm 1, \ldots$, be a bounded strictly increasing sequence of real numbers. Let A be a generating set and define the operator J on $L^2(\Omega, d\mu)$ by $(Jf)(x) = \lambda_n f(x)$ for $x \in A_n$. Then $(Jf)(x) - (UJU^*f)(x) = (\lambda_n - \lambda_{n-1})f(x)$ for x in A_n. Thus $J - UJU^* = D \geq 0$ and 0 is not in the point spectrum of D. Hence by Theorem 2.3.2, U is absolutely continuous. Since U now plays the role of the shift operator considered in § 2.2, it follows that meas $\mathrm{sp}(U) = 2\pi$. Another proof of this last inequality follows from Theorem 2.3.1. For if meas $\mathrm{sp}(U) < 2\pi$, it would follow from that theorem that J is even absolutely continuous, a contradiction.

The portion of Theorem 2.9.2 concerning the measure of $\mathrm{sp}(U)$ was proved in Putnam [21] under an additional hypothesis. That $\mathrm{sp}(U)$ is the entire unit circle whenever T is dissipative also follows from a result of A. Ionescu-Tulcea [1].

§ 2.10 Commutators of self-adjoint operators

Theorem 2.10.1. *Let A, B and C denote self-adjoint operators on a Hilbert space $\mathfrak{H}$ satisfying*

$$AB - BA = iC \text{ on a linear subset } \Omega \text{ of } \mathfrak{H}. \qquad (2.10.1)$$

Suppose that C is bounded and definite (so that either $C \geq \varepsilon I$ or $C \leq -\varepsilon I$ for some $\varepsilon > 0$) and that the set $B(\Omega)$ is dense. Then A must be unbounded.

Proof. The argument is similar to one given in Putnam [4]; cf. also Venkataraman [1]. It will be clear that there is no loss of generality in supposing that C is positive. Suppose, if possible, that A is bounded. Then there exists some λ not belonging to $\mathrm{sp}(A)$. Since, for $x \in \Omega$, $(AB - BA)x = (A_\lambda B - BA_\lambda)x$, where $A_\lambda = A - \lambda I$, it can be supposed that $\lambda = 0$ and so A^{-1} is bounded. Then

$$AB(\Omega) \text{ is dense.} \qquad (2.10.2)$$

Otherwise there would exist an element $y \neq 0$ such that $y \perp AB(\Omega)$. Thus for all $x \in \Omega$, $0 = (y, ABx) = (Ay, Bx)$ and therefore, since $B(\Omega)$ is dense, $Ay = 0$ and so A^{-1} cannot exist, a contradiction.

Since $B(\Omega)$ is dense so is $\mathfrak{R}_B$, and 0 is not in the point spectrum of B. It will be shown that

$$B^{-1} \text{ is bounded} . \qquad (2.10.3)$$

To see this, let $x \in \Omega$, so that by (2.10.1),

$$(ABx, x) - (BAx, x) = i(Cx, x),$$

hence

$$\|ABx\| \geq \text{const.} \|C^{\frac{1}{2}}x\| .$$

For y arbitrary in $\mathfrak{H}$ it follows from (2.10.2) that there exist $x_n \in \Omega$ such that $ABx_n \to y$. Since $\|AB(x_n - x_m)\| \geq \text{const.} \|C^{\frac{1}{2}}(x_n - x_m)\|$ then $C^{\frac{1}{2}}x_n \to z$ for some z in $\mathfrak{H}$. Since $C > 0$ then $C^{-\frac{1}{2}}$ is bounded and therefore $x_n \to x$ ($= C^{-\frac{1}{2}}z$). Since A^{-1} is bounded then $ABx_n \to y$ implies that $Bx_n \to A^{-1}y$. But, since B is self-adjoint, it is closed; hence $x \in \mathfrak{D}_B$ and $Bx = A^{-1}y$. So $\mathfrak{D}_{B^{-1}} = \mathfrak{R}_B \supset \mathfrak{R}_{A^{-1}} = \mathfrak{D}_A = \mathfrak{H}$ and hence (2.10.3) holds.

Next, let $y \in AB(\Omega)$, so that $y = ABx$ with $x \in \Omega$. Therefore $x = B^{-1}A^{-1}y$ and (2.10.1) implies

$$y - BAB^{-1}A^{-1}y = iCB^{-1}A^{-1}y .$$

On applying $A^{-1}B^{-1}$ one gets

$$A^{-1}B^{-1}y - B^{-1}A^{-1}y = iQCQ^*y,$$

where $Q = A^{-1}B^{-1}$, for all y in $AB(\Omega)$. But $AB(\Omega)$ is dense and hence

$$Q - Q^* = iQCQ^* . \tag{2.10.4}$$

If the operator V is defined by $V = C^{\frac{1}{2}}Q^*C^{\frac{1}{2}} + iI$, then it is seen that

$$V^*V = C^{\frac{1}{2}}(QCQ^* + i(Q - Q^*))C^{\frac{1}{2}} + I = I,$$

so that V is isometric. Hence the set W_V is contained in the disk $|z| \leq 1$. Hence $W_{C^{\frac{1}{2}}Q^*C^{\frac{1}{2}}}$ is contained in the circle of unit radius with center at $z = -i$, and $W_{C^{\frac{1}{2}}QC^{\frac{1}{2}}}$ is contained in the circle of unit radius with center at $z = i$. Since $\|C^{\frac{1}{2}}x\| \geq \text{const.} \|x\|$, with const. > 0, it follows that there exists a positive number μ with the property that W_Q is a subset of the circle of radius μ and center at $z = i\mu$, that is,

$$W_Q \subset \{z : |z - i\mu| \leq \mu\} . \tag{2.10.5}$$

Since B^{-1} exists, there exists some number $\lambda \neq 0$ in the spectrum of B^{-1}, hence

$$B^{-1}x_n - \lambda x_n \to 0, \quad \|x_n\| = 1, \lambda \neq 0, \tag{2.10.6}$$

holds for some sequence of unit vectors x_n in $\mathfrak{D}_{B^{-1}}$. But

$$(Q^*x_n, x_n) = (B^{-1}QBx_n, x_n) = (QBx_n, B^{-1}x_n) = (Qx_n, x_n) + \varepsilon_n ,$$

where $\varepsilon_n \to 0$ as $n \to \infty$. This means that $\text{Im}(Qx_n, x_n) \to 0$ and hence, by (2.10.5), $(Qx_n, x_n) \to 0$.

Let $Q = H + iJ$ denote the Cartesian form for Q. Then $(Jx_n, x_n) \to 0$ and, since $J = (1/2i)(Q - Q^*) \geq 0$, also $Jx_n \to 0$. By (2.10.4), $J = \frac{1}{2}QCQ^*$ and so $C^{\frac{1}{2}}Q^*x_n \to 0$, hence $Q^*x_n \to 0$ ($C^{-\frac{1}{2}}$ being bounded). Since also $Jx_n \to 0$ then $Qx_n \to 0$, that is $A^{-1}B^{-1}x_n \to 0$. Since A is bounded, $B^{-1}x_n \to 0$, a contradiction to (2.10.6). This completes the proof of Theorem 2.10.1.

It can be noted that Theorem 1.5.1 implies that if A, B and C satisfy the hypotheses of Theorem 2.10.1 then at least one of the pair A, B must be unbounded. However it is possible that one, say B, is bounded, as will be seen in the examples to be considered below.

§ 2.11 Examples

For a fixed real θ, $0 \leq \theta < 2\pi$, consider the differential operator $A = -id/dt$ on $L^2(0, 1)$ with domain $\mathfrak{D}_A = \{x(t): x(t) \text{ absolutely continuous, } x'(t) \in L^2(0, 1), x(0) = e^{i\theta}x(1)\}$, that is, the set of functions $x(t)$ of the form

$$x(t) = ce^{-i\theta t} + \int_0^t y(s)\,ds\,, \qquad (2.11.1)$$

where

$$c = \text{const.}, \quad y \in L^2(0,1) \quad \text{and} \quad \int_0^1 y(s)\,ds = 0\,;$$

see Stone [1], p. 428, also von Neumann [7], p. 137. For each fixed θ, A_θ is a self-adjoint operator. Let B denote the bounded self-adjoint multiplication operator on $L^2(0, 1)$ defined by $Bx = tx$. Then

$$A_\theta B - B A_\theta = -iI \qquad (2.11.2)$$

holds on the set $\Omega_\theta = \mathfrak{D}_{A_\theta B} \cap \mathfrak{D}_{B A_\theta}$. Since Ω_θ contains all functions of class C^1 which, along with their derivatives, are zero at the end-points, it is clear that $B(\Omega_\theta)$ is dense in $\mathfrak{H} = L^2(0, 1)$. Thus the unboundedness of A would follow from Theorem 2.10.1. Actually this and more is easily demonstrated directly, in fact

$$\text{sp}(A_\theta) = \{2\pi n - \theta\}\,, \qquad n = 0, \pm 1, \pm 2, \dots\,. \qquad (2.11.3)$$

The above example shows however that the conditions of Theorem 2.10.1 can be fulfilled with B bounded. It is noteworthy that, although both Ω_θ and $B(\Omega_\theta)$ are dense, $A_\theta(\Omega_\theta)$ is not, as is seen directly (or as a consequence of Theorem 2.10.1, since otherwise, B would have to be unbounded).

If $\mathfrak{H} = L^2(-\infty, \infty)$ and if $A = -id/dt$ and $Bx = tx$ with

$$\mathfrak{D}_A = \{x \in L^2(-\infty, \infty): x \text{ absolutely continuous, } x' \in L^2(-\infty, \infty)\} \qquad (2.11.4)$$

and

$$\mathfrak{D}_B = \{x \in L^2(-\infty, \infty): tx \in L^2(-\infty, \infty)\}\,, \qquad (2.11.5)$$

then again A and B are self-adjoint (cf. von Neumann [7], Chapt. II §§ 8,9; Stone [1], p. 441) and

$$AB - BA = -iI \quad \text{on} \quad \Omega \equiv \mathfrak{D}_{AB} \cap \mathfrak{D}_{BA}\,. \qquad (2.11.6)$$

In this case it is easily shown that $B(\Omega)$ and $A(\Omega)$ are dense, so that the unboundedness of both A and B would follow from Theorem 2.10.1. Incidentally, both A and B are absolutely continuous operators.

It is noteworthy that the condition that $B(\Omega)$ be dense in Theorem 2.10.1 does not imply, for example, that $\mathrm{sp}(A)=(-\infty,\infty)$ as happens to be the case above. In the preceding example it is seen from (2.11.3) that A_θ has a discrete spectrum. Furthermore, even if $\mathrm{sp}(A)=(-\infty,\infty)$, the condition that $B(\Omega)$ be dense does not imply the absolute continuity of A. In fact, by allowing θ in the first example to run through the rationals on $(0,2\pi)$ and by taking direct sums of Hilbert spaces and corresponding operators A_θ, it is clear that the example could be modified so that, for instance, $\mathrm{sp}(A)=(-\infty,\infty)$ but is a dense pure point spectrum.

The above results show that the condition that $B(\Omega)$ be dense in the hypothesis of Theorem 2.10.1 cannot be replaced by the condition that Ω be dense. (In view of the essentially symmetric roles of A and B in the statement of Theorem 2.10.1 with this modified hypothesis there would follow the false conclusion that both A and B must be unbounded.) Moreover, it can be remarked that the assumption $C>0$ cannot be relaxed to the condition $C\geqq 0$ and 0 not in the point spectrum of C. In fact, it is easy to construct examples where the latter conditions hold with both A and B bounded and $A(\Omega)$ and $B(\Omega)$ dense (in fact A and B can be supposed non-singular and $\Omega=\mathfrak{H}$); see, e.g., § 6.22 on Jacobi matrices.

§ 2.12 More on non-negative perturbations and spectra

It was shown above that if $B=J+D$ where J is bounded from below, D is bounded and non-negative, and if $B=UJU^*$ holds for some unitary U, then certain assertions can be made concerning the measure and absolute continuity of the spectrum of U. In this section some results will be obtained in which the boundedness or half-boundedness will be relaxed or omitted.

For any transformation T and an arbitrary complex number λ let $T_\lambda=T-\lambda I$.

Theorem 2.12.1. *Let B, J and D be self-adjoint operators satisfying*

$$B = J+D \quad \text{with } D \geqq 0.\qquad(2.12.1)$$

(so that, in particular, $Bx=Jx+Dx$ for all x in $\mathfrak{D}_B=\mathfrak{D}_J\cap\mathfrak{D}_D$). Suppose that there exists a closed interval of length $\|D\|$, $0\leqq\|D\|\leqq\infty$, having no points in common with the spectrum of J. Further, let μ be any real number not in the spectrum of B, that is, 0 not in the spectrum of B_μ, and for which the interval $[-\|D\|,0]$ (which, if $D=0$, is the single point 0 and, if D is unbounded, the interval $(-\infty,0]$) has no points in common with $\mathrm{sp}(J_\mu)$. Then there holds the relation

$$J_\mu^{-1}-B_\mu^{-1} = J_\mu^{-1}DB_\mu^{-1} \geqq 0.\qquad(2.12.2)$$

Proof. Choose μ as in the theorem and note that $B_\mu=J_\mu+D$. Since

$\mathfrak{H} = \mathfrak{D}_{B_\mu^{-1}} = \mathfrak{R}_{B_\mu}$ then for any x in $\mathfrak{H}$, $x = B_\mu y$ for some y in $\mathfrak{D}_{B_\mu}$. Consequently,

$$y = B_\mu^{-1} x \in \mathfrak{D}_{B_\mu}$$

and so

$$x = J_\mu B_\mu^{-1} x + D B_\mu^{-1} x,$$

hence

$$J_\mu^{-1} x = B_\mu^{-1} x + J_\mu^{-1} D B_\mu^{-1} x$$

for all x in $\mathfrak{H}$ and the first part of (2.12.2) is proved. (Note that B_μ^{-1}, J_μ^{-1}, hence also $J_\mu^{-1} D B_\mu^{-1}$, are bounded.)

Next, it will be shown that the bounded self-adjoint operator $M = J_\mu^{-1} D B_\mu^{-1}$ is non-negative. First, suppose that 0 is not in the point spectrum of D. Then M^{-1} exists and $M^{-1} = B_\mu D^{-1} J_\mu$ (cf., e.g., Sz.-Nagy [1], p. 28). Let $x \in \mathfrak{D}_{M^{-1}}$. Then, since $D^{-1} J_\mu x \in \mathfrak{D}_{B_\mu}$, one has

$$(M^{-1} x, x) = (B_\mu D^{-1} J_\mu x, x) = ((J_\mu + D) D^{-1} J_\mu x, x)$$
$$= (D^{-1} J_\mu x, J_\mu x) + (J_\mu x, x)$$
$$\geq \|D\|^{-1} \|J_\mu x\|^2 + (J_\mu x, x)$$
$$= \int (\|D\|^{-1} \lambda^2 + \lambda) d\|E_\lambda x\|^2,$$

where $J_\mu = \int \lambda dE_\lambda$ is the spectral resolution of J_μ. However the hypothesis implies that $\|D\|^{-1} \lambda^2 + \lambda > 0$ for $\lambda \in \mathrm{sp}(J_\mu)$ and so $M^{-1} \geq 0$, hence also $M \geq 0$, as was to be shown.

It remains to be shown that $M^{-1} \geq 0$ (hence $M \geq 0$) when 0 belongs to the point spectrum of D. In this case, let $\varepsilon > 0$ and put $D^\varepsilon = D + \varepsilon I$ and $J_{\mu\varepsilon} = J_\mu - \varepsilon I$. Then $B_\mu = J_{\mu\varepsilon} + D^\varepsilon$. Since $D^\varepsilon \geq \varepsilon I$, then 0 is not in $\mathrm{sp}(D^\varepsilon)$. If μ is chosen as before, then, since the spectrum is closed, it is clear that ε can be chosen so small that $[-\|D^\varepsilon\|, 0]$ has no point in common with the spectrum of $J_{\mu\varepsilon}$ or J_μ. Then, as above, $J_{\mu\varepsilon}^{-1} - B_\mu^{-1} = J_{\mu\varepsilon}^{-1} D^\varepsilon B_\mu^{-1} \geq 0$. But it is clear that $J_{\mu\varepsilon}^{-1} \rightrightarrows J_\mu^{-1}$ as $\varepsilon \to 0$ (the convergence being in the uniform norm topology) and hence $J_\mu^{-1} - B_\mu^{-1} \geq 0$. This completes the proof.

In case $J > 0$ (hence $B > 0$) then one can choose $\mu = 0$ in Theorem 2.12.1. In this case the assertion of Theorem 2.12.1 is similar to a result of Rellich [3], p. 363 (but where $B \geq J$ is defined in terms of $B^{\frac{1}{2}}$ and $J^{\frac{1}{2}}$; see also Heinz [1], p. 422) concerning monotone properties of positive operators. In fact, these results are generalizations of the monotone property of bounded positive operators: $B \geq J > 0 \Rightarrow J^{-1} \geq B^{-1}$, which in turn is a generalization of a result of Löwner [1]. In this connection it can be noted that Heinz [1] has obtained generalizations of other (related) results of Löwner on monotonicity of operators. See also Beckenbach and Bellman [1], Bellman [1], Bendat and Sherman [1], Wigner and Yanase [1].

In case B and J are unitarily equivalent, so that

$$B = UJU^* \text{ for some unitary } U, \tag{2.12.3}$$

then $B_\mu^{-1} = UJ_\mu^{-1}U^*$ (or $J_\mu^{-1} = U^*B_\mu^{-1}U$) if μ is defined as in Theorem 2.12.1. It follows that the results of §§ 2.2 and 2.3 can be applied with the appropriate change of notation for the self-adjoint operators occurring there. For instance, Theorems 2.3.1, 2.3.2 and 2.12.1 imply the following.

Theorem 2.12.2. *Let B, J and D be self-adjoint operators satisfying (2.12.1) and (2.12.3). Let μ be any real number satisfying the conditions of Theorem 2.12.1 (note that $\mathrm{sp}(J) = \mathrm{sp}(B)$). Then $\mathfrak{H}_a(U)$ contains the least subspace of $\mathfrak{H}$ reducing U and containing the range of the bounded operator $J_\mu^{-1}DB_\mu^{-1}$. If, in addition,*

$$\mathrm{meas}\ \mathrm{sp}(U) < 2\pi, \tag{2.12.4}$$

then both $\mathfrak{H}_a(U)$ and $\mathfrak{H}_a(J)$ contain the least subspace of $\mathfrak{H}$ reducing U and J and containing the range of $J_\mu^{-1}DB_\mu^{-1}$.

Remark. The above result clearly implies the Corollary of Theorem 2.8.1.

§ 2.13 Commutators of self-adjoint operators

Again let $T_\lambda \equiv T - \lambda I$.

Theorem 2.13.1. *Let A, J and C be self-adjoint, $C \geq 0$, and suppose that*
$$AJ - JA = -iC \text{ on } \Omega = A_i^{-1}(\mathfrak{D}_J \cap \mathfrak{D}_{A^\dagger - 1 CA_i^{-1}}), \tag{2.13.1}$$

(so that $AJx - JAx = -iCx$ for x in Ω). If J is bounded from below, then there exists a real μ, not belonging to $\mathrm{sp}(J)$, and a bounded self-adjoint operator $S = J_\mu^{-1} A_i^{*-1} CA_i^{-1} B_\mu^{-1} \geq 0$, where $B = UJU^*$ and $U = A_i A_i^{*-1}$ is the Cayley transform of A, such that

$$\mathfrak{H}_a(A) \supset \mathfrak{R}_S. \tag{2.13.2}$$

In particular, if 0 is not in the point spectrum of C then A is absolutely continuous.

If also
$$\mathrm{sp}(A) \neq (-\infty, \infty), \tag{2.13.3}$$

then both $\mathfrak{H}_a(A)$ and $\mathfrak{H}_a(J)$ contain the least subspace of $\mathfrak{H}$ reducing A and J and containing $\mathfrak{R}_S$.

Proof. It is clear that $A_i J - JA_i = -iC$ holds on Ω. If x is in $A_i(\Omega)$ then $x = A_i y$, y in Ω, and so

$$A_i JA_i^{-1}x - Jx = -iCA_i^{-1}x,$$

therefore

$$JA_i^{-1}x - A_i^{-1}Jx = -iA_i^{-1}CA_i^{-1}x$$

for all x in $A_i(\Omega)$. Hence

$$J(I + 2iA_i^{-1})x - (I + 2iA_i^{-1})Jx = 2A_i^{-1}CA_i^{-1}x.$$

If U denotes the Cayley transform of A,

$$U = A_i A_i^{*-1} = I - 2iA_i^{*-1},$$

then $U^* = I + 2iA_i^{-1}$ and so $JU^*x - U^*Jx = 2A_i^{-1}CA_i^{-1}x$. Hence

$$UJU^* = J + 2A_i^{*-1}CA_i^{-1} \tag{2.13.4}$$

on $A_i(\Omega)$. But $A_i(\Omega) = \mathfrak{D}_J \cap \mathfrak{D}_{A_i^{*-1}CA_i^{-1}}$ and so (2.13.4) holds as an operator equation. It is clear that $A_i^{*-1}CA_i^{-1} \geqq 0$. Since J is bounded from below, a number μ as in Theorem 2.12.2 exists (for instance, $\mu = $ const. $< \min J$) and relation (2.13.2) then follows from Theorem 2.12.2 if it is noted that $\mathfrak{H}_a(U) = \mathfrak{H}_a(A)$. Moreover, if also (2.13.3) is assumed, then (2.12.4) holds for the present U, and the assertions concerning $\mathfrak{H}_a(A)$ and $\mathfrak{H}_a(J)$ also follow from Theorem 2.12.2. This completes the proof.

A result similar to the above is the following.

Theorem 2.13.2. *Let A, J and C be self-adjoint, $C \geqq 0$, and suppose also that C is bounded. Suppose that*

$$AJ - JA = -iC \quad \text{on} \quad \Omega_\lambda = (\lambda A - iI)^{-1}(\mathfrak{D}_J) \tag{2.13.5}$$

for all sufficiently small $\lambda > 0$ (or even for a sequence $\lambda_n > 0$, $\lambda_n \to 0$). If

$$\mathrm{sp}\,(J) \neq (-\infty, \infty), \tag{2.13.6}$$

then there exists a real μ not in $\mathrm{sp}\,(J)$ and, for some $\alpha > 0$, a bounded self-adjoint operator $S_\alpha = J_\mu^{-1}(\alpha A + iI)^{-1}C(\alpha A - iI)^{-1}B_\mu^{-1} \geqq 0$, where $B = U_\alpha J U_\alpha^*$ and U_α is the Cayley transform of αA, such that

$$\mathfrak{H}_a(A) \supset \mathfrak{R}_{S_\alpha}. \tag{2.13.7}$$

In particular, if 0 is not in the point spectrum of C then A is absolutely continuous.

If also (2.13.3) holds then $\mathfrak{H}_a(A)$ and $\mathfrak{H}_a(J)$ contain the least subspace of $\mathfrak{H}$ reducing A and J and containing $\mathfrak{R}_{S_\alpha}$.

Proof. In view of (2.13.6) and the boundedness of C, one can choose $\alpha > 0$ so small that there exists a closed interval of length $2\alpha \|C\|$ in the complement of $\mathrm{sp}\,(J)$. Moreover it can be assumed that (2.13.5) holds with $\lambda = \alpha$. Then $(\alpha A)J - J(\alpha A) = i\alpha C$ holds on Ω_α. If U_α is the Cayley transform of αA, so that $U_\alpha = (\alpha A - iI)(\alpha A + iI)^{-1}$, then corresponding to (2.13.4) one has

$$U_\alpha J U_\alpha^* = J + 2\alpha Q_\alpha^* C Q_\alpha \tag{2.13.8}$$

on $(\alpha A - iI)\Omega_\alpha \equiv D_J$, where $Q_\alpha = (\alpha A - iI)^{-1}$. Thus (2.13.8) holds as an operator equation. (Note that $Q_\alpha^* C Q_\alpha$ is bounded.) Clearly $Q_\alpha^* C Q_\alpha \geqq 0$ and, since $\|Q_\alpha\| \leqq 1$, also $\|2\alpha Q_\alpha^* C Q_\alpha\| \leqq 2\alpha \|C\|$. Hence, in view of the choice of α, there exists a closed interval of length $\|2\alpha Q_\alpha^* C Q_\alpha\|$ in the complement of sp (J). It is clear that $\mathfrak{H}_a(U_\alpha)=\mathfrak{H}_a(A)$ and (2.13.7) follows from Theorem 2.12.2. If (2.13.3) holds then clearly meas sp $(U_\alpha)<2\pi$ and the proof is completed by another application of Theorem 2.12.2.

An immediate consequence is the following.

Corollary. *Let A, J and C be self-adjoint, let C be bounded and non-negative, and suppose that (2.13.5) holds for sufficiently small $\lambda>0$. In addition, suppose that 0 is not in the point spectrum of C and that sp $(J)\neq (-\infty,\infty)$. Then A is absolutely continuous. If also sp $(A)\neq(-\infty,\infty)$, then J is also absolutely continuous.*

As an example, let $A=-id/dt$ and $J=V(t)$, where $V(t)$ is a real-valued, bounded function of class C^1 on $(-\infty,\infty)$ with a bounded, positive derivative $V'(t)$. Let the domain of A be given by (2.11.4). Then A and J are self-adjoint operators on $L^2(-\infty,\infty)$, and, if $C=V'(t)$, $AJ-JA=-iC$ holds on $\mathfrak{D}_A$. Clearly J and C are bounded, $C\geqq 0$, and 0 is not in the point spectrum of C. Since

$$\Omega = (\lambda A-iI)^{-1}(\mathfrak{D}_J)=(\lambda A-iI)^{-1}(\mathfrak{H})=\mathfrak{D}_A,$$

relation (2.13.5) holds for any $\lambda>0$. It follows from the above Corollary that A is absolutely continuous. (This fact is well-known and can also be established directly, since A is unitarily equivalent via the Fourier transform to the coordinate operator; cf. § 4.2.) It is clear that J is also absolutely continuous although this does not follow from the Corollary.

§ 2.14 An application to quantum mechanics

Let A_0 and H denote self-adjoint operators, let $\{U_t\}$ be a unitary group given by $U_t=e^{itH}$ for $-\infty<t<\infty$, and define A_t by

$$A_t = U_t A_0 U_t^*, \quad U_t = e^{itH}. \tag{2.14.1}$$

If A_t denotes an observable at time t associated with a quantum mechanical system with Hamiltonian H, and if A_t and H do not depend explicitly on t, then, for a proper normalization of constants, the evolutionary dependence of A_t upon A_0 is given in the Heisenberg representation by (2.14.1); see, e.g., Ludwig [1], p. 111, Kramers [1], p. 160, Mackey [1], p. 84.

In certain cases the existence of an absolutely continuous part of H, which may reduce to H itself, can be inferred from the behavior of special observables A_0. There will be proved the following result (cf. Putnam [26]).

Theorem 2.14.1. *Let A_0 and H be self-adjoint and suppose that A_t is defined by (2.14.1). In addition, suppose that D is self-adjoint and satisfies, for some real t,*

$$A_t = A_0+D, \text{ where } D\geqq 0. \tag{2.14.2}$$

Then

(i) $\mathfrak{H}_a(H) \supset \mathfrak{R}_D$ *if* A_0 *is bounded*;

(ii) $\mathfrak{H}_a(H) \supset \mathfrak{R}_{(A_0 - \mu I)^{-1} D(A_t - \mu I)^{-1}}$ *if* $A_0 \geq kI$ *and* $\mu < k$.

In particular it follows that if A_0 is bounded from below and if (2.14.2) holds with a D having an inverse (so that 0 is not in the point spectrum of D), then H is absolutely continuous.

Proof. First, let A_0 be bounded. Then D is also bounded and $D = U_t(A_0 U_t^*) - (A_0 U_t^*)U_t$. If H has the spectral resolution $H \doteq \int \lambda \, dE_\lambda$, so that $U_t = \int e^{it\lambda} dE_\lambda$, the argument used in the proof of Theorem 2.3.2 then yields $DE(Z) = 0$ (hence $E(Z)D = 0$) for any Borel set Z of measure 0, and so (i) follows.

Next, let A_0 be half-bounded and satisfy $A_0 \geq kI$. If $\mu < k$ it follows from Theorem 2.12.1 that

$$(A_0 - \mu I)^{-1} - (A_t - \mu I)^{-1} = (A_0 - \mu I)^{-1} D(A_t - \mu I)^{-1} \geq 0.$$

Since $(A_t - \mu I)^{-1} = U_t(A_0 - \mu I)^{-1} U_t^*$ and $(A_0 - \mu I)^{-1}$ is bounded, the argument used to prove (i) now yields (ii) and the proof of Theorem 2.14.1 is complete.

Consider the following example. Suppose that A_0 is an arbitrary self-adjoint operator and (2.14.1) holds. In addition suppose that $dA_t/dt = I$, that is, that $A_t = A_0 + tI$ for $-\infty < t < \infty$. (See Ludwig [1], p. 110, for a specific quantum mechanical example.) It is clear from the uniqueness of the spectral resolution that if $\{F_\lambda\}$ is the spectral family of A_0 then the spectral family $\{F_{t\lambda}\}$ of A_t is given by $F_{t\lambda} = F_{\lambda - t}$. It follows from (2.14.1) that $F_{\lambda - t} = U_t F_\lambda U_t^*$. Since $F_{\lambda - t} - F_\lambda$ is semi-definite, relation (i) of Theorem 2.14.1 then implies that $\mathfrak{H}_a(H)$ contains the range of $F_{\lambda - t} - F_\lambda$ for all λ and t on $(-\infty, \infty)$, and hence H must be absolutely continuous.

Formally, relation (2.14.1) is equivalent to the commutation relation

$$dA/dt = i(HA - AH). \tag{2.14.3}$$

The actual equivalence exists of course under certain restrictions, e.g., if A_0 and H are bounded. In any case, however, the results of § 2.13 could be applied to (2.14.3).

Chapter III

Semi-normal operators

§ 3.1 Introduction

As in § 1.5, a bounded operator T will be called semi-normal if

$$TT^* - T^*T = D, \quad D \geq 0 \text{ or } D \leq 0. \tag{3.1.1}$$

Clearly any normal operator is also semi-normal. It is easy to see that the converse is also true in case the space $\mathfrak{H}$ is finite-dimensional. For if, say, $D \geq 0$, its eigenvalues are non-negative while their sum is the trace of D, which is 0. Hence all eigenvalues are 0 and so $D = 0$. In the infinite-dimensional case however it is possible that an operator be semi-normal without being normal. In fact, any isometric but not unitary operator V has this property; for $V^*V = I$ and $VV^* - V^*V \leq 0, \neq 0$. On l^2 the operator A given by the matrix $A = (a_{ij})$ with $a_{i+1,i} = 1$ and $a_{ij} = 0$ otherwise $(i, j = 1, 2, \ldots)$ is such an operator.

Remark. The term semi-normal as used here was introduced by Halmos [4], p. 237. The concept itself however was considered by Halmos in his earlier paper [2] (cf. p. 129), where an operator T satisfying (3.1.1) with $D \leq 0$ was called subnormal. Today this latter term has a somewhat different meaning; see § 3.14. If in (3.1.1), D satisfies $D \leq 0$, T is also said to be hyponormal, a term introduced by Berberian [2], p. 161.

The present chapter will be concerned mainly with the spectra of a semi-normal operator and of its real and imaginary parts.

§ 3.2 Structure properties

The following was proved in Putnam [27].

Theorem 3.2.1. *Let T be a (bounded) semi-normal operator on the Hilbert space $\mathfrak{H}$, so that (3.1.1) holds. Let $\mathfrak{M} = \mathfrak{M}_T$ denote the smallest subspace of $\mathfrak{H}$ reducing T and containing the range of D and let $T = H + iJ$ denote the Cartesian form of T. Then*

$$\mathfrak{H}_a(H) \supset \mathfrak{M}; \tag{3.2.1}$$

and, if $\mathfrak{M}^{\perp}$ denotes the orthogonal complement of $\mathfrak{M}$ (so that $\mathfrak{M}^{\perp}$ also reduces T) then

$$T \text{ is normal on } \mathfrak{M}^{\perp}. \tag{3.2.2}$$

In addition,

$$\|D\| \leq (2/\pi)\|J\| \text{ meas sp}(H), \tag{3.2.3}$$

and the inequality (3.2.3) is optimal in the sense that there exist bounded operators T satisfying (3.1.1) with $D \neq 0$ for which (3.2.3) becomes an equality.

Proof. A simple calculation shows that $HJ - JH = iC$ where $D = 2C$. Relation (3.2.3) follows from (2.2.1) of Theorem 2.2.1. The assertion (3.2.1) follows from Theorem 2.2.4 if it is noted that the space $\mathfrak{L}$ occurring there is now the least space reducing both H and J (that is, reducing T) and containing the range of C (that is, the range of D). Assertion (3.2.2) is a consequence of the fact that $\mathfrak{M}^{\perp}$ is clearly contained in the null space of D. Finally, the assertion concerning the optimal nature of the constants can be deduced by examples involving Hilbert transforms; see § 6.12 below.

If θ is real it is clear that $e^{i\theta} T$ is also semi-normal whenever T is, so that the assertions of Theorem 3.2.1 hold if $T = H + iJ$ is replaced by $e^{i\theta} T = H_\theta + iJ_\theta$, with

$$H_\theta = \tfrac{1}{2}(e^{i\theta} T + e^{-i\theta} T^*) \quad \text{and} \quad J_\theta = (1/2i)(e^{i\theta} T - e^{-i\theta} T^*).$$

Corresponding to (3.2.1) and (3.2.3) one can conclude, if $\theta = -\pi/2$, that

$$\mathfrak{H}_a(J) \supset \mathfrak{M} \tag{3.2.4}$$

and

$$\|D\| \leq (2/\pi)\|H\| \text{ meas sp}(J). \tag{3.2.5}$$

Further, since sp(H) and sp(J) must have positive measure whenever T is semi-normal but not normal, it is clear that if T is semi-normal and completely continuous then (since $T + T^*$ is also completely continuous) T must be normal. See Andô [1], Berberian [3], Stampfli [1], also Putnam [27], p. 818.

Although the preceding theorem implies that the spectrum of the real (or imaginary) part of a semi-normal, non-normal operator has positive measure, it apparently is an open question whether it must always contain an interval. Is it possible, for instance, that a Cantor set (nowhere dense perfect set) of positive measure can be the spectrum of the real part of a semi-normal operator?

§ 3.3 Spectrum of a semi-normal operator

Let T be semi-normal, so that (3.1.1) holds. In case T is not normal it will be shown that under certain conditions the spectrum of T has a positive two-dimensional measure. It is however an open question whether the spectrum of every semi-normal, non-normal operator has positive (two-dimensional) measure.

Let $T_\theta = e^{i\theta} T$ for θ real and let

$$H_\theta = \tfrac{1}{2}(T_\theta + T_\theta^*). \tag{3.3.1}$$

It is seen that H_θ is the real or imaginary part of T according as $\theta = 0$ or $\theta = -\pi/2$. If $\lambda \in \mathrm{sp}(T)$, then λ will be called accessible if there exists a sequence $\{\lambda_n\}$, $\lambda_n \notin \mathrm{sp}(T)$, satisfying $\lambda_n \to \lambda$ as $n \to \infty$.

The next theorem is due to Putnam [11].

Theorem 3.3.1. *Let T be bounded and satisfy $TT^* - T^*T = D \geq 0$ and let $\lambda = re^{-i\theta}$ $(r \geq 0)$ be an accessible point of* $\mathrm{sp}(T)$. *Then*

$$(\max H_\theta)^2 \geq \min TT^* \tag{3.3.2}$$

and

$$|r - \max H_\theta| \leq [(\max H_\theta)^2 - \min TT^*]^{\frac{1}{2}}. \tag{3.3.3}$$

Proof. Let $\lambda_n = r_n e^{-i\theta_n}$ be chosen so that $\lambda_n \notin \mathrm{sp}(T)$ and $\lambda_n \to \lambda$ as $n \to \infty$. Put $T_n = T - \lambda_n I$. Since $T_n T_n^* = T_n T_n^* T_n T_n^{-1}$ then $\mathrm{sp}(T_n T_n^*) = \mathrm{sp}(T_n^* T_n)$ and hence $\mathrm{sp}(TT^* - 2rH_\theta) = \mathrm{sp}(T^*T - 2rH_\theta)$. (Cf. Putnam and Wintner [1], p. 76, footnote.) Since $\lambda = re^{-i\theta} \in \mathrm{sp}(T)$, then either $(T - \lambda I)x_m \to 0$ or $(T^* - \bar{\lambda}I)x_m \to 0$ for some sequence of unit vectors x_m. In either case it follows from $TT^* - T^*T \geq 0$ that

$$\lim_{m \to \infty} \sup (x_m, T^*Tx_m) \leq r^2$$

and that

$$\lim_{m,n \to \infty} (x_m, H_{\theta_n} x_m) = r.$$

Hence $\min(TT^* - 2rH_\theta) \leq -r^2$ and so $\min(TT^*) - 2r \max(H_\theta) + r^2 \leq 0$. Relations (3.3.2) and (3.3.3) now follow.

Corollary 1. *Let T be bounded and satisfy $TT^* - T^*T \geq 0$ and suppose that $0 \in \mathrm{sp}(T)$ and $\min(TT^*) > 0$. Then, for all θ, the disk $|\lambda| \leq \max H_\theta - [(\max H_\theta)^2 - \min(TT^*)]^{\frac{1}{2}}$ lies in* $\mathrm{sp}(T)$.

Proof. One need only note that $\lambda = 0$ is in $\mathrm{sp}(T)$ but by (3.3.3) no accessible points of $\mathrm{sp}(T)$ can lie in the interior of the disk specified. Hence the entire disk lies in $\mathrm{sp}(T)$.

Even for an arbitrary bounded operator T, not necessarily semi-normal, if $0 \in \mathrm{sp}(T)$ and $\min TT^* > 0$ then there exists some disk $|\lambda| \leq$ const. belonging to $\mathrm{sp}(T)$ (see, e.g., Putnam and Wintner [1], pp. 76–78). But if $TT^* - T^*T \geq 0$ the radius specified in Corollary 1 can be claimed.

Corollary 2. *If V is isometric and not unitary then its spectrum is the disk $|\lambda| \leq 1$.*

Proof. This result is well-known and can be deduced from a normal form for isometric operators; cf., e.g., Stone [1], pp. 351 ff. The proof as a consequence of Theorem 3.3.1 is as follows. Let $T^* = V$ so that $TT^* = I$ and $TT^* - T^*T \geq 0$. Since $\text{sp}(T^*) = \overline{\text{sp}(T)}$ and since $\min(TT^*) = (\max H_\theta)^2 = 1$, the assertion follows from Corollary 1.

The following was proved by Andô [1], Berberian [3], Stampfli [1].

Theorem 3.3.2. *If T is semi-normal then* $\| T^n \| = \| T \|^n$ *for* $n = 1, 2, \ldots$

Proof. The proof below is that of Andô [1]. It can be supposed that $TT^* - T^*T \leq 0$ so that $\| T^*x \| \leq \| Tx \|$ for all $x \in \mathfrak{H}$. It is sufficient to show that if $\| T \| = 1$ then $\| T^n \| = 1$ for all n. This in turn will certainly be established if it is shown that (when $\| T \| = 1$) for every $\varepsilon > 0$, and for each $n = 0, 1, 2, \ldots$, there exists a unit vector x such that

$$\| T^n x \| \geq 1 - \varepsilon \quad \text{and} \quad \| T^n x - T^* T^{n+1} x \| \leq \varepsilon . \tag{3.3.4_n}$$

The validity of $(3.3.4_0)$ is clear, since $1 \in \text{sp}(T^*T)$. Suppose the relation valid for n; its validity for $n+1$ will be shown. One has

$$\| T^{n+1} x - T^* T^{n+2} x \|^2 = \| T^{n+1} x \|^2 - 2 \| T^{n+2} x \|^2 + \| T^* T^{n+2} x \|^2$$
$$\leq \| T^n x \|^2 - \| T^{n+2} x \|^2$$
$$\leq \| T^n x \|^2 - \| T^* T^{n+1} x \|^2$$
$$\leq \| T^n x - T^* T^{n+1} x \| (\| T^n x \|$$
$$+ \| T^* T^{n+1} x \|) \leq 2\varepsilon .$$

But

$$\| T^{n+1} x \| \geq \| T^{n+2} x \| \geq \| T^* T^{n+1} x \| \geq \| T^n x \| - \varepsilon \geq 1 - 2\varepsilon ,$$

and, since $\varepsilon > 0$ is arbitrary, relation $(3.3.4_{n+1})$ follows.

It is seen that if T is semi-normal its spectral radius

$$r_T = \lim_{n \to \infty} \| T^n \|^{1/n} = \| T \| ,$$

a well-known property of normal operators. It can be mentioned that the last theorem implies that if T is semi-normal, but not normal, then T is not completely continuous. (See the remark following formula line (3.2.5) above.)

Berberian [3] has also shown that if T is semi-normal with an invariant subspace $\mathfrak{M}$ and if the restriction of T to $\mathfrak{M}$ is completely continuous then $\mathfrak{M}$ reduces T.

Further similarities between semi-normal and normal operators will be discussed in the next section.

§ 3.4 Further spectral properties

The material of §§ 3.4–3.7 is based mostly on Putnam [32].

If T satisfies

$$TT^* - T^*T = D \quad \text{with} \quad D \geq 0 \text{ or } D \leq 0 , \tag{3.4.1}$$

and if $T = H + iJ$ is the Cartesian representation of T then as noted earlier,

$$HJ - JH = iC, \quad \text{where} \quad D = 2C \geqq 0 \text{ or } \leqq 0. \qquad (3.4.2)$$

It will be shown that the spectra of the real and imaginary parts respectively of a semi-normal operator T are precisely the projections of the spectrum of T onto the real and imaginary axes. This result is known for normal operators and can be deduced, for instance, from the spectral resolution formula for T. There will be proved the following result.

Theorem 3.4.1. Let T satisfy (3.4.1). (i) If $\lambda_0 \in \mathrm{sp}(H)$ there exists some real λ_0' and a sequence $\{x_n\}$ of unit vectors for which $(H - \lambda_0 I)x_n \to 0$ and $(J - \lambda_0' I)x_n \to 0$ as $n \to \infty$, so that in particular, $\lambda_0 + i\lambda_0' \in \mathrm{sp}(T)$. Similarly, if $\mu_0 \in \mathrm{sp}(J)$, there exists some real μ_0' and a sequence $\{y_n\}$ of unit vectors for which $(H - \mu_0' I)y_n \to 0$ and $(J - \mu_0 I)y_n \to 0$ as $n \to \infty$, so that, in particular, $\mu_0' + i\mu_0 \in \mathrm{sp}(T)$. (ii) If λ_0 and μ_0 are real and if $\lambda_0 + i\mu_0 \in \mathrm{sp}(T)$ then $\lambda_0 \in \mathrm{sp}(H)$ and $\mu_0 \in \mathrm{sp}(J)$.

Proof of (i). Since iT is also semi-normal and has the Cartesian form $iT = (-J) + iH$, it is clearly sufficient to prove only the first part of (i). It will be clear from the proof that there is no loss of generality in supposing that $D \geq 0$.

Let $\lambda_0 \in \mathrm{sp}(H)$. Then there exists a sequence $\{f_n\}$ of unit vectors satisfying $(H - \lambda_0 I)f_n \to 0$, hence also $J(H - \lambda_0 I)f_n \to 0$. By (3.4.2),

$$(H - \lambda_0 I)J - J(H - \lambda_0 I) = iC$$

and so

$$(f_n, (H - \lambda_0 I)Jf_n) - (f_n, J(H - \lambda_0 I)f_n) = i \| C^{\frac{1}{2}} f_n \|^2 \to 0.$$

Hence $Cf_n \to 0$ and so $(H - \lambda_0 I)Jf_n \to 0$. If now Jf_n is identified with the previous f_n, then $(H - \lambda_0 I)J^2 f_n \to 0$ and, similarly, $(H - \lambda_0 I)p(J)f_n \to 0$ where $p(J)$ is any polynomial in J. If $\phi(\lambda)$ is any continuous function on $-\infty < \lambda < \infty$ and if $\phi(J)$ is defined by the usual functional calculus, then $\phi(J)$ can be uniformly approximated by polynomials in J and so

$$(H - \lambda_0 I)\phi(J)f_n \to 0, \quad \| f_n \| = 1. \qquad (3.4.3)$$

Let J have the spectral resolution $J = \int \lambda \, dF_\lambda$ and suppose that $\mathrm{sp}(J)$ is contained in the interior of $\Delta_1 = [c, d]$. Then $\| F(\Delta_1)f_n \| = 1$ for all n. Clearly, for at least one of the two intervals $\Delta = [c, \frac{1}{2}(c+d)]$ or $\Delta = [\frac{1}{2}(c+d), d]$, say $\Delta = \Delta_2$, there holds $\| F(\Delta_2)f_n^{(2)} \| \geq \frac{1}{2}$ $(n = 1, 2, \ldots)$, where $\{f_n^{(2)}\}$ is a subsequence of $\{f_n^{(1)}\}$, with $f_n^{(1)} = f_n$. Continuation of this process leads to a sequence of intervals $\Delta_1, \Delta_2, \ldots$, where

$$\Delta_{k+1} \subset \Delta_k, \quad |\Delta_k| = (d-c)/2^{k-1},$$

$\{f_n^{(k+1)}\}$ is a subsequence of $\{f_n^{(k)}\}$ and $\| F(\Delta_k)f_n^{(k)} \| \geq 1/2^{k-1}$ $(k, n = 1, 2, \ldots)$.

Let λ_0' denote the real number determined by the nested sequence $\{\Delta_k\}$, so that $c_k, d_k \to \lambda_0'$ as $k \to \infty$, where $\Delta_k = [c_k, d_k]$. For each $k = 1, 2, \ldots$, choose $\gamma_k > 0$ so that $\gamma_k \to 0$ as $k \to \infty$ and define the continuous function $\phi_k(\lambda)$ on $-\infty < \lambda < \infty$ as the function the graph of which consists of the real axis from $-\infty$ to $(c_k - \gamma_k, 0)$, the three segments joining $(c_k - \gamma_k, 0)$ to $(c_k, 1)$ to $(d_k, 1)$ to $(d_k + \gamma_k, 0)$, and the real axis from $(d_k + \gamma_k, 0)$ to ∞.

It is clear that $0 < 1/2^{k-1} \leq \|F(\Delta_k)f_n^{(k)}\| \leq \|\phi_k(J)f_n^{(k)}\| \leq \|f_n^{(k)}\|$. If $g_{kn} = \phi_k(J)f_n^{(k)}/\|\phi_k(J)f_n^{(k)}\|$, then $\|g_{kn}\| = 1$ and, by (3.4.3), $(H - \lambda_0 I)g_{kn} \to 0$ as $n \to \infty$ (for each fixed k). On the other hand it is clear that $\|(J - \lambda_0' I)g_{kn}\| \leq d_k - c_k + \gamma_k$ (all n). Hence there exists a subsequence $\{n_k\}$ of the positive integers with the property that if $x_k = g_{kn_k}$, then both $(H - \lambda_0 I)x_k \to 0$ and $(J - \lambda_0' I)x_k \to 0$ as $k \to \infty$, as was to be shown.

Proof of (ii). Let $\nu_0 = \lambda_0 + i\mu_0 \in \mathrm{sp}(T)$. It will be shown that $\lambda_0 \in \mathrm{sp}(H)$. (A similar argument shows that $\mu_0 \in \mathrm{sp}(J)$.) Again it can be supposed that $D \geq 0$. If $T_\nu = T - \nu I$ with $\nu = \lambda + i\mu$, then a simple calculation shows that

$$T_\nu T_\nu^* = (H - \lambda I)^2 + (J - \mu I)^2 + C. \tag{3.4.4}$$

If $T_{\nu_0} T_{\nu_0}^*$ is singular, there exists a sequence $\{x_n\}$ of unit vectors satisfying $(T_{\nu_0} T_{\nu_0}^* x_n, x_n) \to 0$. Since $C \geq 0$ this implies by (3.4.4) that $(H - \lambda_0 I)x_n \to 0$ and $(J - \mu_0 I)x_n \to 0$ and (ii) is proved. In case $T_{\nu_0} T_{\nu_0}^* > 0$ then necessarily $T_{\nu_0}^* T_{\nu_0}$ is singular and hence there exists a disk about ν_0 lying in $\mathrm{sp}(T)$ (see § 3.3). Let μ_0' be the maximum value of μ with the property that, for $\nu = \lambda_0 + i\mu$, T_ν is singular. Clearly, $\nu_0' = \lambda_0 + i\mu_0' \in \mathrm{sp}(T)$ and $T_{\nu_0'} T_{\nu_0'}^*$ must be singular. As before it follows that $\lambda_0 \in \mathrm{sp}(H)$ (also $\mu_0' \in \mathrm{sp}(J)$) and the proof of (ii) is complete.

An immediate consequence of Theorem 3.4.1 is the following.

Corollary 1. *If T is semi-normal and if its spectrum is real then T is self-adjoint.*

Both the preceding corollary and the next were given by Putnam [32] and Stampfli [3]. The next corollary was also proved by Saitô and Yoshino [2]. See also Yoshino [1].

Corollary 2. *If T is semi-normal then the set W_T is the smallest closed convex set containing $\mathrm{sp}(T)$.*

Proof of Corollary 2. Note that for a self-adjoint operator A the set W_A is a closed segment of the real axis joining the maximum and minimum points of $\mathrm{sp}(A)$. Also, if θ is real, then $T_\theta = e^{i\theta} T$ is also semi-normal. Since $\mathrm{sp}(T_\theta) = e^{i\theta}\mathrm{sp}(T)$ and $W_T = e^{i\theta} W_T$ it follows from Theorem 3.4.1 that W_T is contained in every closed rectangle of the complex plane which contains $\mathrm{sp}(T)$. Hence W_T is contained in the intersection of all such rectangles, that is, W_T is contained in the smallest closed convex set containing $\mathrm{sp}(T)$. Since, even for arbitrary T, $\mathrm{sp}(T)$ is a subset of W_T, the proof is complete.

It is clear that Corollary 2 implies Corollary 1 since, for an arbitrary bounded operator A, the reality of the set W_A implies the self-adjointness of A.

Another proof of Corollary 2 above can be deduced from Theorem 3.10.2 below. See § 3.10, also Stampfli [3].

In case T is normal the assertion of Corollary 2 is well-known and is due to Toeplitz. For a proof in this special case, see, e.g., Stone [1] or, for a recent proof, see Berberian [5]. For some related results see Halmos [10], Hildebrandt [1–3], Meng [1], Schreiber [5], Stampfli [6].

It follows from the proof of Theorem 3.4.1 above that if T is semi-normal and if some power T^m is completely continuous then so also is T (and hence, as noted in §§ 3.2, 3.3, T must be normal). In order to see this, suppose if possible that $T = H + iJ$ is not completely continuous and hence either H or J is not completely continuous. It can be supposed that H is not completely continuous, so that there exists a number $\lambda_0 \neq 0$ in the essential spectrum of H. Consequently, by Weyl's criterion (Weyl [1], cf. Riesz and Sz.-Nagy [1], p. 364), $(H - \lambda_0 I)f_n \to 0$ holds for a sequence of unit vectors $\{f_n\}$ converging weakly to 0. If the present vectors f_n are identified with those occurring in the proof of (i) of Theorem 3.4.1 above, it is seen that for each fixed k the sequence $\{g_{kn}\}$ occurring there converges weakly to 0 as $n \to \infty$. Choose the index n_k so large that both

$$\|(H - \lambda_0 I)g_{kn_k}\| < 1/k \quad \text{and} \quad |(g_{kn_k}, g_{ij})| < 1/k, \text{ for } i,j = 1, \ldots, k,$$

hold. It is seen that if $x_k = g_{kn_k}$ then $(x_k, x) \to 0$ as $k \to \infty$ for each $x = g_{ij}$ (i,j fixed) and hence also for each x in the closure of the linear manifold determined by the g_{ij}'s. Since a similar relation obviously holds for x in the orthogonal complement of this space, it is clear that $\{x_k\}$ converges weakly to 0. As before, $(T - v_0 I)x_k \to 0$ as $k \to \infty$, where $v_0 = \lambda_0 + i\lambda_0'$, and consequently $(T^m - v_0^m I)x_k \to 0$ as $k \to \infty$ for each fixed $m = 1, 2, \ldots$. Since $v_0^m \neq 0$ then T^m is not completely continuous for each $m = 1, 2, \ldots$, a contradiction.

Remark. In general a bounded operator A may have a completely continuous power even though A itself is not completely continuous. In fact, if A is the operator on the one-sided sequential Hilbert space l^2 obtained by taking the direct sum of the 2×2 matrices $\begin{pmatrix} 0 & 1 \\ 0 & 0 \end{pmatrix}$ then $A^2 = 0$ (and hence is completely continuous). If e_n denotes the vector in l^2 with n-th component 1 and all others 0, then $\{e_n\}$ converges weakly to 0 but $\|Ae_n\| = 1$ for $n = 2, 4, 6, \ldots$, and so A is not completely continuous.

§ 3.5 An integral formula

For an arbitrary bounded operator T on a Hilbert space define the function $M(x)$ on $-\infty < x < \infty$ by

$$M(x) = \begin{cases} \sup \text{Im}(z) - \inf \text{Im}(z), & z \in \text{sp}(T), x = \text{Re}(z) \\ 0 \text{ if } x \notin \text{Re}(\text{sp}(T)). \end{cases} \quad (3.5.1)$$

Thus $M(x)$ is the distance between the upper and lower boundaries of $\text{sp}(T)$ over the point x. If T_θ is defined by

$$T_\theta = e^{i\theta} T, \quad (3.5.2)$$

let $M_\theta(x)$ correspond to T_θ as $M(x)$ $(=M_0(x))$ does to $T(=T_0)$.

Theorem 3.5.1. *Let $T = H + iJ$ be semi-normal, so that*

$$TT^* - T^*T = D \text{ where } D \geq 0 \text{ or } D \leq 0. \quad (3.5.3)$$

Then for every real θ,

$$\pi\|D\| \leq \int_{-\infty}^{\infty} M_\theta(x)\,dx. \quad (3.5.4)$$

More generally, if $\text{Re}(T_\theta) = \frac{1}{2}(T_\theta + T_\theta^*)$ has the spectral resolution

$$\text{Re}(T_\theta) = \int \lambda \, dE_\lambda^\theta, \quad (3.5.5)$$

and if S denotes any Borel set of the real axis, then

$$\pi\|E^\theta(S)DE^\theta(S)\| \leq \int_S M_\theta(x)\,dx. \quad (3.5.6)$$

Since $E^\theta(S) = I$ in case $S = (-\infty, \infty)$ it is clear that (3.5.4) follows from (3.5.6). Further, in case S has measure 0, it follows from (3.5.6) and the semi-definiteness of D that $DE^\theta(S) = 0$, a result proved earlier (see §§ 2.2, 2.4).

Proof. Let $H = \text{Re}(T)$ have the spectral resolution $H = \int \lambda \, dE_\lambda$. If $\varDelta = (a, b]$ is any half-open interval, and if $A_\varDelta = E(\varDelta)AE(\varDelta)$, then clearly $A_\varDelta$ leaves invariant the Hilbert space $\mathfrak{H}_\varDelta = E(\varDelta)\mathfrak{H}E(\varDelta)$. Let $\text{sp}(A_\varDelta)$ denote the spectrum of $A_\varDelta$ as an operator on $\mathfrak{H}_\varDelta$.

It follows from (3.5.3) (cf. (3.4.2)) that if $J = \text{Im}(T)$ then $H_\varDelta J_\varDelta - J_\varDelta H_\varDelta = iC_\varDelta$, where $C = \frac{1}{2}D$ and $C_\varDelta \geq 0$ or $C_\varDelta \leq 0$ according as $C \geq 0$ or $C \leq 0$. Thus $T_\varDelta$ is semi-normal on $\mathfrak{H}_\varDelta$. For arbitrary μ, one has

$$(H_\varDelta - \mu I_\varDelta)J_\varDelta - J_\varDelta(H_\varDelta - \mu I_\varDelta) = iC_\varDelta.$$

If μ is chosen to be the midpoint of $\varDelta$ then $\|H_\varDelta - \mu I_\varDelta\| \leq \frac{1}{2}|\varDelta|$ and hence by a relation similar to (3.2.5),

$$(2\pi)^{\frac{1}{2}}\|C^{\frac{1}{2}}E(\varDelta)x\| \leq [|\varDelta| \, \text{meas}_1 (\text{sp}(J_\varDelta))]^{\frac{1}{2}}\|E(\varDelta)x\|, \quad (3.5.7)$$

for all $x \in \mathfrak{H}$, where meas_1 refers to ordinary one-dimensional Lebesgue measure. (It can be assumed here that $C \geq 0$.)

Next, let $\mu_0 \in \text{sp}(J_\varDelta)$. Then by Theorem 3.4.1 there exists some real number λ_0 and a sequence $\{x_n\}$ of unit vectors in $\mathfrak{H}_\varDelta$, thus $\|x_n\| = 1$ and $x_n = E(\varDelta)x_n$, for which

$$(H - \lambda_0 I)x_n \to 0 \tag{3.5.8}$$

and

$$E(\Delta)(J - \mu_0 I)x_n \to 0. \tag{3.5.9}$$

(Note that $E(\Delta)HE(\Delta) = HE(\Delta)$ and that $\lambda_0 \in \mathrm{sp}\,(H_\Delta)$, so that $\lambda_0 \in \Delta'$, the closure of Δ.) Relation (3.5.9) implies that

$$((J - \mu_0 I)x_n, x_n) \to 0, \tag{3.5.10}$$

and hence there exist real numbers μ_1 and μ_2 satisfying

$$\mu_1 \leqq \mu_0 \leqq \mu_2; \ \mu_1, \mu_2 \in \mathrm{sp}\,(J). \tag{3.5.11}$$

Next, it will be shown that there exists a point $\mu_1' \leqq \mu_0$ for which $\lambda_0 + i\mu_1' \in \mathrm{sp}\,(T)$. To this end, note that if $(J - \mu_0 I)x_n \to 0$ as $n \to \infty$, then one can choose $\mu_1' = \mu_0$. Hence, it can be supposed that

$$\lim_{n \to \infty} \sup \|(J - \mu_0 I)x_n\| > 0. \tag{3.5.12}$$

As in § 3.4, suppose that $\mathrm{sp}\,(J)$ is contained in the interior of $[c, d]$ so that $c < \mu_0$. If $\Delta_1 = [c, \mu_0]$ it follows from (3.5.10) and (3.5.12) that

$$\lim_{n \to \infty} \sup \|F(\Delta_1)(J - \mu_0 I)x_n\| > 0,$$

where $J = \int \lambda \, dF_\lambda$. Hence there exists a subsequence $\{y_n\}$ of $\{x_n\}$ for which both $(H - \lambda_0 I)y_n \to 0$ and $\|F(\Delta_1)y_n\| > \mathrm{const.} > 0$.

The argument of § 3.4 can now be applied to yield a point $\mu_1' \in \Delta_1$, hence $\mu_1' \leqq \mu_0$, and a sequence $\{z_n\}$ of unit vectors for which $(H - \lambda_0 I)z_n \to 0$, $(J - \mu_1' I)z_n \to 0$ as $n \to \infty$. (The present Δ_1 plays the role of $[c, d]$ in the argument of § 3.4.) Hence $\lambda_0 + i\mu_1' \in \mathrm{sp}\,(T)$; a similar argument shows that $\lambda_0 + i\mu_2' \in \mathrm{sp}\,(T)$ for some $\mu_2' \geqq \mu_0$.

Consequently, when $\mu_0 \in \mathrm{sp}\,(J_\Delta)$ there exist a real number $\lambda_0 \in \Delta'$, the closure of Δ, and a pair μ_1', μ_2' satisfying $\mu_1' \leqq \mu_0 \leqq \mu_2'$ such that $\lambda_0 + i\mu_1', \lambda_0 + i\mu_2' \in \mathrm{sp}\,(T)$. Hence $\mathrm{meas}_1\,(\mathrm{sp}\,(J_\Delta)) \leqq Q(\Delta')$, where $Q(\delta)$ denotes the interval function defined by $Q(\delta) = 0$ if $\delta \cap \mathrm{Re}\,(\mathrm{sp}\,(T))$ is empty and $Q(\delta) = \sup \mathrm{Im}\,(z) - \inf \mathrm{Im}\,(z)$ where $z \in \mathrm{sp}\,(T)$ and $\mathrm{Re}\,(z) \in \delta$.

Relation (3.5.7) now yields

$$(2\pi)^{\frac{1}{2}} \|C^{\frac{1}{2}} E(\Delta)x\| \leqq [|\Delta| Q(\Delta')]^{\frac{1}{2}} \|E(\Delta)x\|. \tag{3.5.13}$$

If $(c, d] \supset \mathrm{sp}\,(H)$ and if $P: c = c_0 < c_1 < \ldots < c_N = d$ is a partition of $(c, d]$ into subintervals $\Delta_k = (c_{k-1}, c_k]$ then

$$I = \sum_{k=1}^N E(\Delta_k) \quad \text{and} \quad \|C^{\frac{1}{2}}x\| = \|C^{\frac{1}{2}} \sum_{k=1}^N E(\Delta_k)x\| \leqq \sum_{k=1}^N \|C^{\frac{1}{2}} E(\Delta_k)x\|.$$

Since

$$\|x\|^2 = \sum_{k=1}^N \|E(\Delta_k)x\|^2,$$

an application of the Schwarz inequality to (3.5.13) then implies

$$(2\pi)^{\frac{1}{2}} \| C^{\frac{1}{2}} x \| \leq \left(\sum_{k=1}^{N} d_k Q(\varDelta'_k) \right)^{\frac{1}{2}} \| x \| ,$$

where d_k is the length of $\varDelta_k$. If $F(x)$ is defined on $(c, d]$ by $F(x) = Q(\varDelta'_k)$ on $\varDelta_k$, then one obtains

$$(2\pi)^{\frac{1}{2}} \| C^{\frac{1}{2}} x \| \leq \left(\int_c^d F(x) \, dx \right)^{\frac{1}{2}} \| x \| . \qquad (3.5.14)$$

Next, choose a sequence of partitions $\{P_n\}$ with the property that P_{n+1} is a refinement of P_n and such that the lengths of the intervals of P_n tend to 0 as $n \to \infty$. Let $F_n(x)$ correspond to P_n as $F(x)$ does to P. Since $\mathrm{sp}(T)$ is a closed set it is clear from the definition of $M(x)$ $(=M_0(x))$ that $Q(\varDelta^n) \to M(x)$ as $n \to \infty$ whenever $\{\varDelta^n\}$ is a sequence of intervals containing x for which $|\varDelta^n| \to 0$ as $n \to \infty$. Thus $F_n(x) \to M(x)$ as $n \to \infty$ for all x on $(c, d]$ except possibly for those x in the (denumerable) set of partitioning points. Since $0 \leq F_n(x) \leq \mathrm{const.}$, it follows from (3.5.14) and Lebesgue's term by term integration theorem that (3.5.4) holds with $\theta = 0$. Since the same argument applies to T_θ, relation (3.5.4) must hold for any real θ.

Finally by considering coverings of any Borel set S by denumerable unions of the type $\cup \varDelta_k$ where $\varDelta_k = (a_k, b_k]$ and the $\varDelta_k$ are pairwise disjoint, a similar argument leads to

$$(2\pi)^{\frac{1}{2}} \| C^{\frac{1}{2}} E(S) x \| \leq \left(\int_S M_\theta(x) \, dx \right)^{\frac{1}{2}} \| x \| . \qquad (3.5.15)$$

Since for any operator A, $\| A \|^2 = \| A^* A \|$ and since $E(S) C E(S) = (C^{\frac{1}{2}} E(S))^* (C^{\frac{1}{2}} E(S))$, relations (3.5.15) and $D = 2C$ yield (3.5.6). This completes the proof of Theorem 3.5.1.

Corollary 1. *Let T be semi-normal, so that $TT^* - T^* T = D$ satisfies $D \geq 0$ or $D \leq 0$. Then*

$$\pi \| D \| \leq \mathrm{meas}_2 (W_T) , \qquad (3.5.16)$$

the measure being ordinary two-dimensional Lebesgue measure.

Proof. Let θ be fixed. Since $\mathrm{sp}(T) \subset W_T$ it is clear from the definition of M_θ that $M_\theta(x)$ is not greater than the distance between the upper and lower boundaries of W_T over the point x of the real axis. Thus the right side of (3.5.4) is not greater than the measure of W_T and (3.5.16) follows from Theorem 3.5.1.

Corollary 2. *Let T be semi-normal, so that $TT^* - T^* T = D$ satisfies $D \geq 0$ or $D \leq 0$. Suppose that for some real θ the set $e^{i\theta} \mathrm{sp}(T)$ has the property that, except possibly for a set of real values x of measure 0, the set $S_x = \{z : z \in e^{i\theta} \mathrm{sp}(T), \mathrm{Re}(z) = x\}$ is either a closed interval, a single point, or the empty set. Then*

$$\pi \| D \| \leq \mathrm{meas}_2 \mathrm{sp}(T) . \qquad (3.5.17)$$

Proof. The proof follows from (3.5.4) if it is noted that $\mathrm{meas}_2\,(e^{i\theta}\,\mathrm{sp}(T))$ $=\mathrm{meas}_2\,\mathrm{sp}\,(T)$ and that in the present case, for almost all x, $M_\theta(x)=$ $\mathrm{meas}_1\,(S_x)$ for $x\in\mathrm{Re}\,(e^{i\theta}\,\mathrm{sp}\,(T))$ and $M_\theta(x)=0$ otherwise.

The hypothesis of the preceding Corollary restricts $\mathrm{sp}(T)$ by requiring the existence of some line L with the property that almost all sections of $\mathrm{sp}(T)$ with lines parallel to L should be intervals or points.

The inequalities of (3.5.16) and (3.5.17) are optimal in the sense that there exist non-normal semi-normal operators T for which both relations become equalities. In fact, if T is isometric but not unitary then $\|D\|=1$ and both $\mathrm{sp}\,(T)$ and W_T are the closed disk $|z|\leqq 1$.

Whether (3.5.17) holds for all semi-normal operators is not known, as is also the question as to whether even

$$\mathrm{meas}_2\,\mathrm{sp}\,(T)>0 \qquad\qquad (3.5.18)$$

must hold for semi-normal, but not normal, operators. The validity of (3.5.18) in certain special cases was shown above; see also §§ 3.7, 3.8, 6.3.

§ 3.6 Isolated parts of sp(T)

For any bounded operator T, by an isolated part σ of $\mathrm{sp}\,(T)$ is meant a subset of $\mathrm{sp}\,(T)$ which lies at a positive distance from its complementary part $\mathrm{sp}\,(T)-\sigma$; see Riesz and Sz.-Nagy [1], pp. 418 ff. It is known that if σ is an isolated part of $\mathrm{sp}\,(T)$ then there exists a "parallel projection" $P=P_\sigma$, a bounded, not necessarily self-adjoint, operator satisfying $P^2=P$ and such that both $P\mathfrak{H}$ and $(I-P)\mathfrak{H}$ are subspaces invariant under T. In addition, $\mathrm{sp}\,(T')=\sigma$, where $T'=T/P\mathfrak{H}$ denotes the restriction of T to the space $P\mathfrak{H}$. In case T is hyponormal, so also is T'; cf. Berberian [2], p. 161, problem 10.

In case T is semi-normal, so that $TT^*-T^*T=D$ is semi-definite, let Ω denote the least subspace of $\mathfrak{H}$ reducing T and containing the range of D. Equivalently, the orthogonal complement $\Omega^\perp$ of Ω is the largest subspace of $\mathfrak{H}$ reducing T and contained in the null space of D, that is, $\Omega^\perp$ is the largest subspace reducing T on which T is normal. It will be supposed that T is not normal, so that $\Omega\neq 0$. The next theorem concerns the operator T on Ω and it can therefore be supposed that $\Omega=\mathfrak{H}$.

Theorem 3.6.1. *Let T be hyponormal, let Ω be defined as above, and suppose that $\mathfrak{H}=\Omega\,(\neq 0)$, so that there do not exist non-trivial subspaces reducing T on which T is normal. For each real θ, let T_θ be defined by $T_\theta=e^{i\theta}T$ and let $M_\theta(x)$ be defined as in the beginning of § 3.5. (i) If S is a Borel set on the real axis, then*

$$M_\theta(x)=0 \ \ a.e. \ \ on \ \ S\Rightarrow E^\theta(S)=0\,, \qquad\qquad (3.6.1)$$

where $\{E_\lambda^\theta\}$ is the spectral family belonging to $\mathrm{Re}(T_\theta)$. (ii) Let σ be an isolated part of sp(*T*) with the parallel projection P and let $T' = T/P\mathfrak{H}$ denote the restriction of *T* to $P\mathfrak{H}$. If $T'_\theta = e^{i\theta} T'$, if $E_\lambda^{\prime\theta}$ is the spectral family belonging to $\mathrm{Re}(T'_\theta)$, and if $M'_\theta(x)$ corresponds to T' as $M_\theta(x)$ does to *T*, then

$$M'_\theta(x) = 0 \quad \text{a.e. on} \quad S \Rightarrow E'^\theta(S) = 0 . \tag{3.6.2}$$

Proof of (i). It can be supposed that $\theta = 0$, since a similar argument with *T* replaced by T_θ takes care of the general case. Relation (3.5.6) implies that if $M_0(x) = 0$ a.e. on *S* then $|C|^{\frac{1}{2}} E(S) = 0$ $(D = 2C)$ and hence

$$CE(S) = 0 , \tag{3.6.3}$$

where $T = H + iJ$ and $\{E_\lambda\}$ is the spectral family belonging to *H*. Since $HJ - JH = iC$ then $H^2 J - HJH = iHC$ and $HJH - JH^2 = iCH$ and so

$$H^2 J - JH^2 = i(HC + CH) . \tag{3.6.4}$$

By (3.6.3), $Cx = 0$ whenever $x \in \mathfrak{H}_S \equiv E(S)\mathfrak{H}$ and, since $\mathfrak{H}_S$ is invariant under *H*, then (3.6.4) implies that $H^2 Jx = JH^2 x$ for such *x*. Similarly, if $x \in \mathfrak{H}_S$, then $H^n Jx = JH^n x$ and $p(H)Jx = Jp(H)x$ for any polynomial $p(H)$. Hence if $\{p_n(H)\}$ is a sequence of polynomials converging strongly to $E(S)$, then

$$E(S) Jx = JE(S)x , \qquad x \in \mathfrak{H}_S , \tag{3.6.5}$$

and so $\mathfrak{H}_S$ is invariant under both *H* and *J*. Hence *T* is reduced by $\mathfrak{H}_S$ and, by (3.6.3), *T* is normal on $\mathfrak{H}_S$. But $\mathfrak{H} = \Omega$ and so $\mathfrak{H}_S = 0$, that is, $E(S) = 0$, as was to be proved.

Proof of (ii). As above it can be supposed that $\theta = 0$. If T' has the Cartesian decomposition

$$T' = H' + iJ' \qquad (H' = \textstyle\int \lambda \, dE'_\lambda) \tag{3.6.6}$$

on the Hilbert space $\mathfrak{H}' = P\mathfrak{H}$, then as above, since $M'_0(x) = 0$ a.e. on *S*, the hyponormal operator T' is reduced by, and is normal on, $\mathfrak{H}'_S \equiv E'(S)\mathfrak{H}'$. Hence *T* leaves the space $\mathfrak{H}'_S$ invariant and its restriction to this space is normal. Since *T* is hyponormal on $\mathfrak{H}$, *T* is reduced by $\mathfrak{H}'_S$; see Berberian [2], p. 161, problem 9. As before this implies $\mathfrak{H}'_S = 0$, that is, $E'(S) = 0$.

Corollary. Let *T* be hyponormal, suppose that $\mathfrak{H} = \Omega \; (\neq 0)$, and let σ denote any isolated part of sp(*T*). Let *Q* denote any open strip of the complex plane bounded by two parallel lines and such that the set $\sigma \cap Q$ is not empty. Then $\sigma \cap Q$ is not a subset of any set *N* with the following property: for some θ, the strip $Qe^{i\theta}$ is perpendicular to the x-axis, intersects the x-axis in an open interval (α, β) and the set $Ne^{i\theta}$ is given by

$$Ne^{i\theta} = \{(x, f(x)): \; \alpha < x < \beta, \; f(x) \text{ single-valued}\} . \tag{3.6.7}$$

Proof. Suppose if possible that the assertion is false, so that Q is a non-empty subset of some set N of the type described. Since $\sigma = \mathrm{sp}(T')$ (see the beginning of this section) then $\sigma e^{i\theta}$ is the spectrum of $T'_\theta = T' e^{i\theta}$, while $\sigma e^{i\theta} \cap Q e^{i\theta}$ is not empty and is a subset of $N e^{i\theta}$. But this last fact and (3.6.7) imply that $M'_\theta(x) = 0$ on (α, β) and so, by (3.6.2), $E'_\theta((\alpha, \beta)) = 0$. By Theorem 3.4.1, this implies that $\mathrm{Re}(\mathrm{sp}(T'_\theta)) \cap (\alpha, \beta)$ is empty, in contradiction with $\sigma e^{i\theta} \cap Q e^{i\theta} \neq$ empty set. This completes the proof of the Corollary.

When $\mathfrak{H} = \Omega$, it follows that no isolated part of $\mathrm{sp}(T)$ is contained in a segment (see Corollary 1 of Theorem 3.4.1) or, for instance, in a proper subset of the boundary of a rectangle or a circle. Although the possibility that an isolated part of $\mathrm{sp}(T)$ might consist of the entire boundary of a rectangle or circle is not ruled out by the above results, such a situation has been eliminated by results of Stampfli [3] (see § 3.11 below).

§ 3.7 Measure of sp(T)

As was pointed out earlier it is not known whether the spectrum of a semi-normal, non-normal, operator T must have positive two-dimensional Lebesgue measure. However there will be proved the following curious result.

Theorem 3.7.1. *Let T be semi-normal and not normal and let $T = H + iJ$ denote its Cartesian representation. Then either*

$$\text{each of the sets } \mathrm{sp}(H) \text{ and } \mathrm{sp}(J) \text{ contains an interval} \qquad (3.7.1)$$

or

$$\pi \|D\| \leq \mathrm{meas}_2 \, \mathrm{sp}(T). \qquad (3.7.2)$$

Proof. Consider the argument of § 3.5 and, in particular, the relations (3.5.7)–(3.5.9). Suppose that neither end-point of $\Delta = (a, b]$ belongs to $\mathrm{sp}(H)$. It will be shown that (3.5.9) can be sharpened to

$$(J - \mu_0 I) x_n \to 0. \qquad (3.7.3)$$

Since $\lambda_0 \in \mathrm{sp}(H_\Delta)$ then $\lambda_0 \in \Delta'$, the closure of Δ. Since the end-points of Δ do not belong to $\mathrm{sp}(H)$, then λ_0 is an interior point of Δ. It follows from (3.5.8) (see the argument leading to (3.4.3)) that if $y_n = (J - \mu_0 I) x_n$ then

$$(H - \lambda_0 I) y_n \to 0. \qquad (3.7.4)$$

Since λ_0 is in the interior of Δ, relations (3.7.4) and (3.5.9) imply that $y_n \to 0$, that is (3.7.3). (That λ_0 be an interior point of Δ is crucial here.) It then follows from (3.5.8) and (3.7.3) that $v_0 = \lambda_0 + i\mu_0 \in \mathrm{sp}(T)$.

Hence, whenever $\mu_0 \in \mathrm{sp}(J_\Delta)$ and the end-points of Δ do not belong to $\mathrm{sp}(H)$, there exists some $v_0 \in \mathrm{sp}(T)$ with $\mathrm{Im}(v_0) = \mu_0$ and $\mathrm{Re}(v_0) = \lambda_0 \in \Delta$.

If the Δ-strip: $\{\lambda + i\mu : \lambda \in \Delta\}$ is subdivided into a finite or infinite number of rectangles by horizontal segments, it is clear that $|\Delta|$ meas$_1$ sp(J_Δ) is not greater than any sum, S_Δ, of the areas of those rectangles containing points of sp(T). It follows from (3.5.7) that

$$(2\pi)^{\frac{1}{2}} \| C^{\frac{1}{2}} E(\Delta)x \| \leq S_\Delta^{\frac{1}{2}} \| E(\Delta)x \| . \tag{3.7.5}$$

In order to prove the theorem suppose now that (3.7.1) does not hold. Clearly it can be supposed that sp(H) does not contain an interval. Let sp(H) belong to the interior of $(\alpha, \beta]$, hence by Theorem 3.4.1, Re$($sp$(T))$ lies in the interior of $(\alpha, \beta]$. In particular, $\alpha, \beta \notin$ sp(H). Consider subdivisions of $(\alpha, \beta]$: $(\alpha, \beta] = \cup \Delta_k$ of a finite or of an infinite (denumerable) union of disjoint subintervals Δ_k of the type Δ and for which no end-points of Δ_k lie in sp(H). Then a relation like (3.7.5) holds with Δ replaced by Δ_k and so, by the Schwarz inequality, $(2\pi)^{\frac{1}{2}} \| C^{\frac{1}{2}} x \| \leq (\Sigma S_{\Delta_k})^{\frac{1}{2}} \| x \|$, thus

$$2\pi \| C \| \leq \Sigma S_{\Delta_k} . \tag{3.7.6}$$

Since sp(H) contains no interval it is clear that meas$_2$ sp$(T) = \inf_{P, S_{\Delta_k}} \Sigma S_{\Delta_k}$, where only those partitions P, finite or infinite, are allowed with points not in sp(H). Relation (3.7.6) now implies (3.7.2) and the proof is complete.

The next theorem is similar to the above.

Theorem 3.7.2. *Let T be semi-normal, let Ω be defined as in the beginning of § 3.6, and suppose that $\mathfrak{H} = \Omega$ ($\neq 0$). If*

$$\text{meas}_2 \text{ sp} (T) = 0 , \tag{3.7.7}$$

then there exist two open sets whose closures are respectively the sets sp(H) *and* sp(J).

Proof. It is sufficient to prove the assertion concerning H. To this end, it will be shown that the set consisting of the union of all open intervals contained in sp(H) is dense in sp(H).

If the assertion were false, there would exist some closed interval k containing a point of sp(H) in its interior, so that

$$E(k) \neq 0 , \tag{3.7.8}$$

with the property that no subinterval of k belongs to sp(H).

It follows from the argument given earlier in this section that

$$(2\pi)^{\frac{1}{2}} \| |C|^{\frac{1}{2}} E(k)x \| \leq [\text{meas}_2 \{z : z \in \text{sp} (T) \text{ and } \text{Re} (z) \in k\}]^{\frac{1}{2}} \| E(k)x \| .$$

However, the assumption (3.7.7) implies that the right side of this last inequality is 0 and hence $CE(k) = 0$. The argument of § 3.6 (cf. (3.6.3) and following) can be used to show that T is reduced by, and is normal on, $E(k)\mathfrak{H}$. Since $\mathfrak{H}=\Omega$, then $E(k)=0$, in contradiction with (3.7.8), and the proof is complete.

As noted earlier, it is conceivable that the assertion of Theorem 3.7.2 is vacuous in the sense that the hypotheses cannot be fulfilled.

§ 3.8 Zero measure of sp(T) and normality

A bounded operator T is said to be a scalar operator if $T=PNP^{-1}$ where P is a positive self-adjoint operator and N is normal; cf. Dunford [2]. The following two theorems are due to Stampfli [1].

Theorem 3.8.1. *If T is a semi-normal scalar operator, say $T=PNP^{-1}$, and if its spectrum has two-dimensional zero measure, then T is normal.*

Proof. It can be supposed that $TT^*-T^*T\leqq 0$, and hence also $(T-zI)(T-zI)^*-(T-zI)^*(T-zI)\leqq 0$ for any complex number z. In particular $\|(T^*-\bar{z}I)x\| \leqq \|(T-zI)x\|$ for any number z and any vector x. Let $N=\int z\,dK_z$ denote the spectral resolution of N. Then, since sp(T) is closed and bounded, for any $\varepsilon>0$ there exists a finite set of half-open, half-closed disjoint squares $R_1,\ldots,R_k$, each of dimension $1/n\times 1/n$, such that

$$\mathrm{sp}(T)\subset \bigcup_{i=1}^{k} R_i \quad\text{and}\quad \text{area}\left(\bigcup_{i=1}^{k} R_i\right) = k/n^2 < \varepsilon\,.$$

If z_i is the center of R_i then for vectors $x_i\in E(R_i)\mathfrak{H}$,

$$\|(N-z_iI)x_i\| = \left(\int_{R_i} |z-z_i|^2\,d\|K_z x_i\|^2\right)^{\frac{1}{2}} \leqq \|x_i\|/n\,.$$

Hence

$$\begin{aligned}
\|(N-z_iI)P^2 x_i\| &= \|N^*-\bar{z}_iI)P^2 x_i\| = \|P(T^*-\bar{z}_iI)Px_i\|\\
&\leqq \|T^*-\bar{z}_iI)Px_i\|\,\|P\| \leqq \|(T-z_iI)Px_i\|\,\|P\|\\
&= \|P(N-z_iI)x_i\|\,\|P\| \leqq \|P\|^2\|x_i\|/n\,.
\end{aligned}$$

Clearly $\|P^2(N-z_iI)x_i\| \leqq \|P\|^2\|x_i\|/n$, and hence

$$\|(NP^2 - P^2N)x_i\| \leqq 2\|P\|^2\|x_i\|/n \tag{3.8.1}$$

whenever $x_i\in E(R_i)\mathfrak{H}$. Since the spaces $\{E(R_i)\mathfrak{H}\}$ are orthogonal and span $\mathfrak{H}$, then for all $y\in\mathfrak{H}$, $\|y\|=1$, one has

$$y = \sum_{i=1}^{k} E(R_i)y \quad\text{with}\quad \Sigma\|E(R_i)y\|^2 = 1\,.$$

It follows from (3.8.1) and the Schwarz inequality that

$$\|(NP^2 - P^2N)y\| \leqq 2\|P\|^2 k^{\frac{1}{2}}/n \leqq 2\|P\|^2 \varepsilon^{\frac{1}{2}}\,.$$

Hence $NP^2 = P^2N$ and, since $P>0$, $NP=PN$, that is $T=N$, and the proof is complete.

Theorem 3.8.2. *If T is semi-normal and if T^n is normal for some positive integer n, then T is normal.*

The proof given by Stampfli [1] is valid for arbitrary n and uses Fuglede's theorem (Theorem 1.6.1) as well as some results of Gonshor [1]. The proof given below will be for $n=2$ only and, although it differs from Stampfli's proof, also uses Fuglede's theorem.

Proof (for $n=2$ only). Let $T=H+iJ$ denote the Cartesian form for T. Then $T^2=(H^2-J^2)+i(HJ+JH)$. Since T^2 is normal and since T^2 commutes with T then T^2 commutes with T^*. Hence T^2 commutes with H (and also J). But $T^2H=HT^2$ leads to

$$-HJ^2+J^2H = i(JH^2-H^2J),$$

hence $HJ^2=J^2H$ and $H^2J=JH^2$. But $HJ-JH=iC$ and so $HC= -CH$. But $HC^2=(HC)C=(-CH)C=C^2H$ and, since $C\geq 0$, $HC=CH$. It now follows from Theorem 1.6.3 that $C=0$, so that $HJ=JH$ and T is normal. (One could also conclude that $C=0$ from the Kleinecke-Sirokov theorem (Theorem 1.3.1) or, since also $JC=CJ$, even from the earlier specialized versions proved by Putnam and Vidav.)

In connection with roots of normal operators, see also Kurepa [1,2,3], Putnam [12], Stampfli [1,2,4].

§ 3.9 Special products of self-adjoint operators

It is easily seen that if T is bounded with the Cartesian form $T= H+iJ$ then T is normal whenever HJ is self-adjoint (and conversely). On the other hand, simple examples show that T need not be normal when HJ is normal. The following is true however.

Theorem 3.9.1. *Let $T=H+iJ$ be semi-normal and suppose that HJ is normal. Then T is normal.*

Proof. In order to see this, note that $Q=HJ$ satisfies $QH=HQ^*$. Since $Q-Q^*=HJ-JH=iC$ with C semi-definite, the theorem II of Beck and Putnam [1] (cf. also Berberian [4]) can be used to conclude that $QH=HQ$, hence $HC=0$ and so $CH=0$. In particular, C commutes with H. As above, this implies $C=0$ as was to be shown.

In case H is non-singular, the proof is easier. For then $Q^*=H^{-1}QH$ so that Q and Q^* have identical spectra. Since $\text{Im}(Q)$ is semi-definite then $\text{sp}(Q)$ is real and hence Q is self-adjoint.

See also the results of Wiegmann [1] and Kaplansky [1] mentioned in § 1.6.

§ 3.10 Resolvents of semi-normal operators

The next two theorems are due to Stampfli [3]. (See also Yoshino [1].)

Theorem 3.10.1. *Let T be hyponormal, so that $T^*T-TT^*\geq 0$ and suppose that $\lambda\notin\text{sp}(T)$. Then $(T-\lambda I)^{-1}$ is hyponormal.*

Proof. It is easily shown that if T is hyponormal so also is $T-\mu I$

for arbitrary μ. Hence it can be supposed that $\lambda=0$. Now
$$T^{-1}T^*TT^{*-1}-I=T^{-1}(T^*T-TT^*)T^{*-1}\geqq 0$$
and so
$$T^*T^{-1}T^{*-1}T=(T^{-1}T^*TT^{*-1})^{-1}\leqq I.$$
Hence
$$T^{*-1}T^{-1}-T^{-1}T^{*-1}=T^{*-1}(I-T^*T^{-1}T^{*-1}T)T^{-1}\geqq 0.$$

Theorem 3.10.2. *If T is hyponormal and if $\lambda\notin\mathrm{sp}(T)$ then*
$$d(\lambda)\|(T-\lambda I)^{-1}x\|\leqq 1, \quad \|x\|=1 \tag{3.10.1}$$
where $d(\lambda)$ denotes the distance from λ to $\mathrm{sp}(T)$.

Proof. By Theorem 3.3.2, the spectral radius of a hyponormal operator T is $\|T\|$. Hence if $\|x\|=1$ then by Theorem 3.10.1,

$$\begin{aligned}
\|(T-\lambda I)^{-1}x\| &\leqq \|(T-\lambda I)^{-1}\| \\
&= \max\{|\mu|:\mu\in\mathrm{sp}((T-\lambda I)^{-1})\} \\
&= [\min\{|\mu|:\mu\in\mathrm{sp}(T-\lambda I)\}]^{-1} \\
&= [\min\{|\mu-\lambda|:\mu\in\mathrm{sp}(T)\}]^{-1}=d^{-1}(\lambda).
\end{aligned}$$

The above theorem furnishes another proof of the fact that when T is semi-normal, the set W_T is the smallest closed convex set containing $\mathrm{sp}(T)$ (see Corollary 2 of Theorem 3.4.1). Since $\mathrm{sp}(T)\subset W_T$, it is sufficient to prove that W_T is contained in the least closed convex set, $\mathscr{C}$, containing $\mathrm{sp}(T)$. Since T is semi-normal implies that $\alpha T+\beta I$ is semi-normal for arbitrary constants α and β, it is sufficient to show that if $\mathscr{C}$ lies on the left of the imaginary axis then so also does the set W_T. That is, it is to be shown that under the above hypothesis on $\mathscr{C}$, if $a+ib\in W_T$ then $a\leqq 0$. But if this were not the case then there would exist an element x, $\|x\|=1$, such that $Tx=(a+ib)x+y$ with $(x,y)=0$ and $a>0$. Hence by Theorem 3.10.2, for $c>0$, $c^2\leqq \|(T-cI)x\|^2=(a-c)^2+b^2+\|y\|^2$ and hence $2ac\leqq a^2+b^2+\|y\|^2$, which is impossible for large c.

§ 3.11 Semi-normal operators and arc spectra

The next result is due to Stampfli [3].

Theorem 3.11.1. *Let T be a semi-normal operator with a spectrum which is a subset of a smooth arc Γ of the complex plane. Then T is normal.*

By a smooth arc is meant here a simple arc or a simple closed curve given by $\Gamma:z=f(s)$, where s is arc length, $a\leqq s\leqq b$, and f is of class C^2.

Proof. An outline only will be given. It can be supposed that T is hyponormal (otherwise replace T by T^*). For $x\in\mathfrak{H}$ and $\lambda\in\rho(T)$, the resolvent set of T, put $R(\lambda,x)=(T-\lambda I)^{-1}x$. A vector valued function $f(\lambda)$ is said to be an analytic extension of $R(\lambda,x)$ if $f(\lambda)$ is defined and analytic on an open set containing $\rho(T)$ and if $(T-\lambda I)f(\lambda)=x$ for λ in this set. Since sp(T) is contained in an arc, $R(\lambda, x)$ possesses only single-valued extensions. By taking the union of all extensions of $R(\lambda, x)$ one obtains a maximal single-valued extension $R_e(\lambda, x)$ of $R(\lambda, x)$. If $\rho(T, x) = \{\lambda : R_e(\lambda, x)$ is analytic at $\lambda\}$ and $\sigma(T, x)$ is the complement of $\rho(T, x)$ (see Dunford [1]), then it can be shown that the hyponormality of T implies that $\sigma(T, x) \supset \overline{\sigma(T^*, x)}$. This relation in turn implies that $(x, y) = 0$ whenever $\sigma(T, x) \cap \sigma(T, y) = 0$. Consequently $\sigma(T, x) \cap \sigma(T, y) = 0$ implies that $\|x\| \leq \|x+y\|$. This boundedness condition together with the growth condition of Theorem 3.10.2 then implies by the theorems 15 and 18 of Dunford [3], that T is a scalar type spectral operator. In particular, since T is a scalar operator and since sp(T) is contained in an arc, sp(T) has zero two-dimensional measure and hence, by Theorem 3.8.1, T must be normal.

It can be noted that Corollary 1 of Theorem 3.4.1 is an immediate consequence of Theorem 3.11.1 as is also the following.

Corollary. *If T is semi-normal and if* sp(T) *lies on the unit circle then T is unitary.*

The above Corollary was also given by Yoshino [1].

For other results concerning operators T satisfying growth or boundedness conditions, see also Donoghue [2], Nieminen [1], Orland [1], Stampfli [3, 5, 7].

§ 3.12 TT^*-T^*T of one-dimensional range

Let P be a one-dimensional projection operator, let H and J be bounded self-adjoint operators satisfying

$$HJ - JH = (i/\pi)P, \tag{3.12.1}$$

and let T be defined by

$$T = H + iJ. \tag{3.12.2}$$

Since $TT^* - T^*T = (2/\pi)P$, T is semi-normal. In addition suppose that the least space containing the range of P and which is invariant under (that is, which reduces) H is the space $\mathfrak{H}$ itself. If $\{E_\lambda\}$, $a \leq \lambda \leq b$, denotes the spectral family of H and if $Pf = e(f, e)$ for some unit vector e in $\mathfrak{R}_P$, let $\sigma(\lambda) = \|E_\lambda e\|^2$.

It was shown by Xa Dao-xeng [1] that H is absolutely continuous. (This result also follows from Theorem 2.2.4. In fact, it is a consequence of relation (2.2.10), which follows from the results of Putnam [9], even if P is replaced by any non-negative operator.) Furthermore, Xa Dao-xeng has proved that $\sigma'(\lambda)$ is essentially bounded and that there exists a real, bounded measurable function $\alpha(t)$ on $[a, b]$ such that T of (3.12.2) is unitarily equivalent to the operator L defined on $L^2(\sigma)$ by

$$(Ly)(t) = ty(t) + i\left[\alpha(t)y(t) + (i\pi)^{-1}\int_a^b \sigma'(s)(s-t)^{-1}y(s)\,ds\right], \tag{3.12.3}$$

where the integral is a Cauchy principal value. If $x(t) = (\sigma'(t))^{\frac{1}{2}} y(t)$ then this last relation readily implies that T is unitarily equivalent to the operator M defined on $L^2(a, b)$ by

$$(Mx)(t) = tx(t) + i \left[\alpha(t) x(t) + (i\pi)^{-1} \int_a^b (\sigma'(s) \sigma'(t))^{\frac{1}{2}} (s-t)^{-1} x(s) \, ds \right].$$
(3.12.4)

A further discussion of singular integral operators similar to that defined in the bracketed part of (3.12.4) will be given in §§ 6.11 ff. See, in particular, § 6.21 and the reference there to Pincus [2].

§ 3.13 An example concerning T^2

It was seen in § 3.10 that the property of hyponormality is preserved under linear and inverse transformations. However the property is not invariant under squaring: thus, T hyponormal does not imply that T^2 is hyponormal. This can be seen from the following example due to Halmos [2].

Let $\mathfrak{K}$ denote an arbitrary Hilbert space and let $\mathfrak{H}$ denote the set of all functions $x = x(n)$ defined on the integers with values in $\mathfrak{K}$ and satisfying $\sum_{-\infty}^{\infty} \|x(n)\|^2 < \infty$. Then $\mathfrak{H}$ becomes a Hilbert space with inner product $(x, y) = \Sigma(x(n), y(n))$. Next, let $\{P_n\}$ be a bounded sequence of non-negative operators on $\mathfrak{K}$, so that $0 \leq P_n \leq (\text{const.})I$, and define the operators U and P on $\mathfrak{H}$ by $Ux(n) = x(n+1)$ and $Px(n) = P_n x(n)$. It is clear that U is unitary and that P is a non-negative bounded operator. Furthermore, if $T = UP$ then $Tx(n) = P_{n+1} x(n+1)$ and $T^* x(n) = P_n x(n-1)$ and hence $T^* Tx(n) = P_n^2 x(n)$ and $TT^* x(n) = P_{n+1}^2 x(n)$. Consequently $T^* T - TT^* \geq 0$ if and only if

$$P_n^2 \geq P_{n+1}^2 \quad \text{for} \quad n = 0, \pm 1, \ldots . \tag{3.13.1}$$

An easy calculation shows that $T^2 x(n) = P_{n+1} P_{n+2} x(n+2)$ and $T^{*2} x(n) = P_n P_{n-1} x(n-2)$ and hence

$$T^{*2} T^2 x(n) = P_n P_{n-1}^2 P_n x(n) \quad \text{and} \quad T^2 T^{*2} x(n) = P_{n+1} P_{n+2}^2 P_{n+1} x(n).$$

Thus T^2 is hyponormal if and only if

$$P_n P_{n-1}^2 P_n \geq P_{n+1} P_{n+2}^2 P_{n+1} \tag{3.13.2}$$

for all n. It will be shown that (3.13.1) does not imply (3.13.2).

To this end, it can be noted that if $\mathfrak{K}$ has dimension ≥ 2, there exist operators a, b satisfying $a \geq b \geq 0$ but for which $a^2 \geq b^2$ does not hold. (In this connection, see § 2.12 and the references cited there concerning monotone properties of operators.) In fact, if $\mathfrak{K}$ is two-dimensional, so that operators on $\mathfrak{K}$ can be regarded as 2×2 matrices, let

$$a = \begin{pmatrix} 2 & 1 \\ 1 & 1 \end{pmatrix} \quad \text{and} \quad b = \begin{pmatrix} 1 & 0 \\ 0 & 0 \end{pmatrix}.$$

Then $a \geqq 0$, $b \geqq 0$ and

$$a - b = \begin{pmatrix} 1 & 1 \\ 1 & 1 \end{pmatrix} \geqq 0 \quad \text{but} \quad a^2 - b^2 = \begin{pmatrix} 4 & 3 \\ 3 & 2 \end{pmatrix}$$

is not semi-definite. Let P_n be the non-negative square root of a for $n \leqq 0$ and the non-negative square root of b for $n > 0$. Then $P_n^2 \geqq P_{n+1}^2$, so that (3.13.1) holds and T is hyponormal. But $P_0 P_{-1}^2 P_0 = a^2$ and $P_1 P_2^2 P_1 = b^2$ so that (3.13.2) fails to hold for $n = 0$. Hence T^2 is not hyponormal.

§ 3.14 Subnormal operators

A bounded operator T on a Hilbert space $\mathfrak{H}$ is said to be subnormal if there exists a Hilbert space $\mathfrak{K}$ containing $\mathfrak{H}$ as a subspace and a normal operator S on $\mathfrak{K}$ such that S leaves $\mathfrak{H}$ invariant and $Tx = Sx$ for all x in $\mathfrak{H}$. Thus, T is the restriction of a normal operator (on $\mathfrak{K}$) to an invariant subspace ($\mathfrak{H}$). (See the remark of § 3.1.)

For example, a subnormal operator which is not normal is the isometric operator T defined by $T : x = (x_0, x_1, x_2, \ldots) \to Tx = (0, x_0, x_1, \ldots)$ on the one-sided sequential Hilbert space ($\mathfrak{H}$). This operator has a unitary extension $U : (Ux_n) = x_{n-1}$ defined on the two-sided sequential Hilbert space ($\mathfrak{H}$).

It is easy to see that every subnormal operator is also hyponormal. For, let T be subnormal on $\mathfrak{H}$ with the normal extension S on $\mathfrak{K}$ (containing $\mathfrak{H}$) and let P be the orthogonal projection of $\mathfrak{K}$ onto $\mathfrak{H}$. For x, y in $\mathfrak{H}$ one has $(T^*x, y) = (x, Ty) = (x, SPy) = (PS^*x, y)$ and so $T^*x = PS^*x$ for all x in $\mathfrak{H}$. Hence, for all x in $\mathfrak{H}$, $\|Tx\| = \|Sx\| = \|S^*x\| \geqq \|PS^*x\| = \|T^*x\|$ and so T is hyponormal.

The example of § 3.13 shows in particular that a hyponormal operator need not be subnormal. In fact, if S were a normal extension of the operator T occurring there, then S^2 would be a normal extension of T^2, so that T^2 would be subnormal, hence hyponormal, a contradiction. See Halmos [2]; also, for another example, Stampfli [1], p. 1458.

A further distinction between hyponormality and subnormality is shown by an example of Stampfli [3] of an operator T all of whose powers T^n $(n = 0, 1, \ldots)$ are hyponormal, but T is not subnormal. That the analogue of Theorem 3.8.2 is false if normal is replaced in both places by subnormal is shown by an example of Stampfli [4]. See these papers (also Stampfli [5, 7]) for some further remarks and results concerning operators which are "almost" normal.

It can be mentioned that the analogue of Theorem 3.8.1 for subnormal operators is valid. That is, if T is subnormal and if sp(T) has zero area then T is normal (Bishop [1]; cf. Stampfli [3], p. 473).

For other results concerning subnormal operators and normal and unitary extensions see, in addition to the papers cited above, also Sz.-Nagy [6] and the survey article of Halmos [9], the latter containing a variety of results, problems, proofs, references, and an entertaining discussion of the subject.

Chapter IV

Commutation relations in quantum mechanics

§ 4.1 Introduction

So far, commutators of the form $AB - BA = -iC$ have occurred in which A and B are self-adjoint and C was either bounded and arbitrary or semi-definite. In this chapter the special case, important in quantum mechanics, in which C is the identity operator will be considered.

Let P and Q denote self-adjoint operators on a Hilbert space $\mathfrak{H}$, let Ω denote a linear subset of $\mathfrak{H}$ satisfying

$$\Omega \subset \mathfrak{D}_{PQ-QP} \quad (= \mathfrak{D}_{PQ} \cap \mathfrak{D}_{QP}) \tag{4.1.1}$$

and suppose that

$$PQ - QP = -iI \quad \text{on } \Omega, \tag{4.1.2}$$

that is, $(PQ - QP)f = -if$ for all f in Ω. What can be said about the operators P and Q? Since (4.1.2) holds for any pair P, Q if the set Ω consists of the zero element only, it is clear that some restrictions requiring that Ω not be too "small" must be imposed. The possibility of the other extreme, namely that $\Omega = \mathfrak{H}$, must necessarily be ruled out. In fact, in this case $\mathfrak{D}_P = \mathfrak{D}_Q = \mathfrak{H}$, so that both P and Q would be bounded, and, as was first proved by Wintner (see Theorem 1.2.1), relation (4.1.2) then cannot hold.

The Schrödinger operators p, q on $L^2(-\infty, \infty)$ are defined by

$$(pf)(x) = -if'(x) \quad ('=d/dx) \quad \text{and} \quad (qf)(x) = xf(x). \tag{4.1.3}$$

Here, for a proper normalization of constants, p is the one-dimensional momentum operator, q is the corresponding coordinate operator, and the relation (4.1.3) is related to the uncertainty principle of quantum mechanics. See, e.g., Ludwig [1], von Neumann [7]. The domain of q is given by

$$\mathfrak{D}_q = \{f \in L^2(-\infty, \infty) : xf \in L^2(-\infty, \infty)\}, \tag{4.1.4}$$

and q is self-adjoint; cf. von Neumann [7], Chapter II, § 9, and Stone [1],

p. 441. In addition, the operator p is also self-adjoint if its domain is given by

$$\mathfrak{D}_p = \{f \in L^2(-\infty, \infty) : f \text{ absolutely continuous} \qquad (4.1.5)$$
$$\text{and } f' \in L^2(-\infty, \infty)\}.$$

For a proof, see von Neumann, Stone, *loc.cit.* It is easily verified that $P = p$ and $Q = q$ determine a solution of (4.1.2) with $\Omega = \mathfrak{D}_{PQ} \cap \mathfrak{D}_{QP}$. Moreover, the pair (p, q) is irreducible in the sense that $\mathfrak{H} = L^2(-\infty, \infty)$ is the only non-trivial space reducing both p and q.

It will be convenient to consider the representations of p, q when $L^2 = L^2(-\infty, \infty)$ is replaced by the sequential space l^2 of vectors $x = (x_0, x_1, \ldots)$ with $\|x\|^2 = \sum_0^\infty |x_j|^2 < \infty$. If $e_0 = (1, 0, \ldots)$, $e_1 = (0, 1, 0, \ldots)$, ..., then $\{e_k\}$ is a complete orthonormal system in l^2. If $\mathfrak{M}$ denotes the linear manifold of finite linear combinations of the e_k's, and if the matrices $A = (a_{ij})$, $B = (b_{ij})$ are defined on $\mathfrak{M}$ by

$$A e_k = -i2^{-\frac{1}{2}}(k^{\frac{1}{2}} e_{k-1} - (k+1)^{\frac{1}{2}} e_{k+1}) \qquad (4.1.6)$$

and

$$B e_k = 2^{-\frac{1}{2}}(k^{\frac{1}{2}} e_{k-1} + (k+1)^{\frac{1}{2}} e_{k+1}) \qquad (e_{-1} = 0),$$

then it is easily shown that A and B are symmetric operators leaving $\mathfrak{M}$ invariant and with the property that $(AB - BA)f = -if$ for $f \in \mathfrak{M}$.

In addition (cf. Theorem 4.5.1 below), the operators A and B are essentially self-adjoint, that is, the least closed extensions $p_0 = A^{**}$ and $q_0 = B^{**}$ are self-adjoint and are the Heisenberg operators. The Heisenberg pair (p_0, q_0) is irreducible on l^2 and corresponds to the Schrödinger pair (p, q) on L^2.

Recall that a system of operators on a Hilbert space $\mathfrak{H}$ is said to be irreducible if the only non-trivial subspace reducing every operator in the set is the space $\mathfrak{H}$ itself.

Henceforth, by a Schrödinger couple (see Foiaş, Gehér and Sz.-Nagy [1]) will be meant any pair (P, Q) of self-adjoint operators on a Hilbert space $\mathfrak{H}$ of dimension $\aleph_0$ for which $P = UpU^*$ and $Q = UqU^*$ for some unitary operator U.

With the above definition one may then consider the following uniqueness problem. Under what conditions can one conclude that any solution pair (P, Q) of (4.1.2) on a Hilbert space $\mathfrak{H}$ is either a Schrödinger couple or the direct sum $(\Sigma \oplus P_\alpha, \Sigma \oplus Q_\alpha)$ on $\mathfrak{H} = \Sigma \oplus \mathfrak{H}_\alpha$ of Schrödinger couples (P_α, Q_α) on $\mathfrak{H}_\alpha$, where the dimension of each $\mathfrak{H}_\alpha$ is $\aleph_0$?

§ 4.2 Unitary groups e^{itP} and e^{isQ}

If P and Q are treated like bounded operators, a formal argument shows that (4.1.2) implies $PQ^n - Q^n P = -i(Q^n)'$, hence $Pf(Q) - f(Q)P =$

$-if'(Q)$ for analytic operators $f(Q)$, where the prime denotes differentiation. If $f(Q) = e^{isQ}$ this yields $e^{-isQ} P e^{isQ} = P + sI$, hence $e^{-isQ} P^n e^{isQ} = (P + sI)^n$ and $e^{-isQ} g(P) e^{isQ} = g(P + sI)$ for analytic g. If $g(P) = e^{itP}$ one obtains, for $-\infty < s, t < \infty$,

$$U_t V_s = e^{its} V_s U_t, \text{ where } U_t = e^{itP} \text{ and } V_s = e^{isQ}. \tag{4.2.1}$$

The above formal argument leading from (4.1.2) to (4.2.1) was given in the quantum mechanical setting by Weyl [3], although the essential calculations are older: see Campbell [1], Hausdorff [1] (see § 1.4), also Baker [1].

If A is any self-adjoint operator with the spectral resolution $A = \int \lambda \, dE_\lambda$, then $U_t = e^{itA} = \int e^{it\lambda} \, dE_\lambda$ is a strongly continuous one-parameter group of unitary operators. Moreover, the infinitesimal generator iA is given by the strong limit

$$iAx = s\text{-}\lim_{h \to 0} h^{-1}(U_h - I)x \; ;$$

see Riesz and Sz.-Nagy [1], pp. 384–385.

If U is the unitary transformation of $L^2(-\infty, \infty)$ onto itself determined by the Fourier-Plancherel transform

$$Uf = (2\pi)^{-\frac{1}{2}} \int_{-\infty}^{\infty} f(y) e^{-iyx} \, dy \; \left(\int_{-\infty}^{\infty} = \text{l.i.m.} \int_a^b \text{ as } a \to -\infty \text{ and } b \to \infty \right)$$

then $p = U^* q U$ (cf. Stone [1], p. 441), from which it easily follows that the corresponding unitary groups $u_t = e^{itp}$ and $v_s = e^{isq}$ are given by

$$(u_t f)(x) = f(x+t), \quad (v_s f)(x) = e^{isx} f(x). \tag{4.2.2}$$

Since, as is easily verified, u_t and v_s satisfy the commutation relation (4.2.1), it follows that all one-parameter groups $U_t = e^{itP}$ and $V_s = e^{isQ}$, which are generated by Schrödinger couples (P, Q) or direct sums of such couples, must satisfy (4.2.1). That in fact all solutions (P, Q) of (4.2.1) are obtained in this way was proved by von Neumann [2], whose result is given in the next section.

§ 4.3 Von Neumann's theorem

Theorem 4.3.1. *If the pair (P, Q) of self-adjoint operators on a Hilbert space $\mathfrak{H}$ is such that the unitary groups $U_t = e^{itP}$ and $V_s = e^{isQ}$ satisfy the commutation relation (4.2.1), then (P, Q) is a Schrödinger couple or the direct sum of such couples.*

It is enough of course to specify the groups $\{U_t\}$ and $\{V_s\}$ along with, say, weak measurability, since in this case it follows from Stone's theorem (cf. Riesz and Sz.-Nagy [1], p. 383; see also Wiener and Wintner [1], the second footnote on p. 812) that the infinitesimal generators (P, Q) are uniquely determined.

Proof. Define the unitary operator $S(t, s)$ by $S(t, s) = e^{-\frac{1}{2}its} U_t V_s$. It is easily verified that

$$S(t, s) S(u, v) = e^{\frac{1}{2}i(tv - su)} S(t+u, s+v) . \tag{4.3.1}$$

Since $S(0, 0) = 1$, then $S(-t, -s) = S(t, s)^{-1}$. If $a(t, s) \in L(K)$, where K denotes the t, s plane, then the operator A defined by

$$(Af, g) = \int\int a(t, s)(S(t, s)f, g) \, dt \, ds \tag{4.3.2}$$

is a bounded operator on $\mathfrak{H}$. Let $a(t, s)$ be called the kernel of A. Straight-forward calculations show that A^* is a similar operator with kernel $a(-t, -s)$ and that if B is another such operator with kernel $b(t, s)$ then so also are $A + B$ and AB with kernels $a(t, s) + b(t, s)$ and

$$\int\int e^{\frac{1}{2}i(tv - su)} a(t-u, s-v) b(u, v) \, du \, dv \tag{4.3.3}$$

respectively.

Next, it will be shown that

$$A = 0 \Rightarrow a(t, s) = 0 \text{ a.e.} \tag{4.3.4}$$

In order to see this note that $A = 0$ implies that $S(-u, -v) A S(u, v) = 0$. Since the latter has the kernel $e^{i(tv - su)} a(t, s)$, it follows that

$$\int\int e^{i(tv - su)} a(t, s)(S(t, s)f, g) \, dt \, ds = 0 \text{ for all } u, v .$$

By standard approximation methods, it follows that $a(t, s)(S(t, s)f, g) = 0$ a.e. for every fixed pair f, g. Hence $a(t, s) S(t, s) = 0$ and hence $a(t, s) = 0$ a.e., that is (4.3.4).

Next, consider the operator A defined by

$$A = \int\int e^{-\frac{1}{4}(t^2 + s^2)} S(t, s) \, dt \, ds , \tag{4.3.5}$$

that is, A is defined by (4.3.2) with $a(t, s) = e^{-\frac{1}{4}(t^2 + s^2)}$. It is clear that A is self-adjoint and that $A \neq 0$. Moreover, it is easily verified that $AS(u, v)A = 2\pi e^{-\frac{1}{4}(u^2 + v^2)} A$. In particular, if $u = v = 0$, it is seen that $A^2 = 2\pi A$ and hence, if $\mathfrak{M} = \{f : Af = 2\pi f\}$ then $\mathfrak{M} = \mathfrak{R}_A$. Furthermore, if f and g belong to $\mathfrak{M}$, so that $Af = 2\pi f$, $Ag = 2\pi g$, a straightforward calculation shows that

$$(S(t, s)f, S(u, v)g) = \exp\left[-\tfrac{1}{4}(t-u)^2 - \tfrac{1}{4}(s-v)^2 + \tfrac{1}{2}i(tv - su)\right](f, g) . \tag{4.3.6}$$

Let $\{\phi_\alpha\}$ be an orthonormal basis for $\mathfrak{M}$, so that $\mathfrak{M} = \text{c.l.m.} \{\phi_\alpha\}$, and let $\mathfrak{H}_\alpha = \text{c.l.m.} \{S(t, s)\phi_\alpha\}$ for $-\infty < t, s < \infty$. It is clear from (4.3.6) that the $\mathfrak{H}_\alpha$ are mutually orthogonal.

Next, let $\mathfrak{H}_1 = \Sigma \oplus \mathfrak{H}_\alpha$ and $\mathfrak{H}_2 = \mathfrak{H}_1^\perp$. It will be shown that $\mathfrak{H}_2 = 0$, so that

$$\mathfrak{H} = \Sigma \oplus \mathfrak{H}_\alpha . \tag{4.3.7}$$

In order to see this, note that by virtue of (4.3.1), each $\mathfrak{H}_\alpha$ is invariant under $S(u, v)$ for u, v arbitrary. Since $S(u, v)^* = S(u, v)^{-1} = S(-u, -v)$, it is clear that each $\mathfrak{H}_\alpha$ is reduced by each $S(u, v)$, and hence $\mathfrak{H}_1$ and $\mathfrak{H}_2$ are reduced by each $S(u, v)$. But $\mathfrak{H}_1 \supset \mathfrak{M} \,(=\mathfrak{R}_A)$ and so $Af = 0$ for $f \in \mathfrak{H}_2$. An argument like that leading to (4.3.4) now implies that $\mathfrak{H}_2 = 0$, hence (4.3.7).

In view of (4.2.1) it is clear that the operators P, Q are reduced by each $\mathfrak{H}_\alpha$. It remains to be shown that

$(P/\mathfrak{H}_\alpha, Q/\mathfrak{H}_\alpha)$ is a (clearly irreducible) Schrödinger couple for each α.

(4.3.8)

For α fixed, let

$$\mathfrak{S} \equiv \mathfrak{H}_\alpha = \text{c.l.m. } \{f_{ts}\}, \quad \text{where } f_{ts} = S(t, s)\phi_\alpha, \tag{4.3.9}$$

hence, by (4.3.1),

$$S(u, v)f_{ts} = e^{\frac{1}{2}i(su - tv)}f_{t+u, s+v} \tag{4.3.10}$$

and, by (4.3.6),

$$(f_{ts}, f_{uv}) = e^{-\frac{1}{4}(t-u)^2 - \frac{1}{4}(s-v)^2 + \frac{1}{2}i(tv - su)}. \tag{4.3.11}$$

It is clear that $U_u = S(u, 0)$, $V_v = S(0, v)$ and, by (4.3.10), that $U_u f_{ts} = e^{\frac{1}{2}isu}f_{t+u, s}$ and $V_v f_{ts} = e^{-\frac{1}{2}itv}f_{t, s+v}$.

Now, for any pair of self-adjoint operators (P, Q) satisfying (4.2.1) it has been shown that $\mathfrak{H}$ can be written as the direct sum of spaces of the form $\mathfrak{S}$ with $S(t, s) = e^{-\frac{1}{2}its}U_t V_s$. It is clear that if (P', Q') is another pair of self-adjoint operators satisfying (4.2.1) then $\mathfrak{H}$ can be written as the direct sum of spaces of the form $\mathfrak{S}'$ where $\mathfrak{S}'$ has properties similar to those of $\mathfrak{S}$ given by (4.3.9)–(4.3.11). In particular if ϕ'_α corresponds to ϕ_α, if $S'(t, s) = e^{-\frac{1}{2}its}U'_t V'_s$ $(U'_t = e^{itP'}, V'_s = e^{isQ'})$ and if $f'_{ts} = S'(t, s)\phi'_\alpha$, the relation (4.3.11) and the corresponding one for the f'_{ts} show that $(f_{ts}, f_{uv}) = (f'_{ts}, f'_{uv})$. If $f'_{ts} = Uf_{ts}$ it is seen that $\|Uf_{ts}\| = \|f_{ts}\|$ and, from (4.3.10), that $U^*S'(u, v)Uf_{ts} = S(u, v)f_{ts}$. Clearly the isometry can be extended in the natural way to the whole of $\mathfrak{S}$ and $\mathfrak{S}'$. Since $U_t = S(t, 0)$, $V_s = S(0, s)$, $U'_t = S'(t, 0)$, $V'_s = S'(0, s)$, it is clear that

$$U^*(P'/\mathfrak{S}')U = P/\mathfrak{S} \quad \text{and} \quad U^*(Q'/\mathfrak{S}')U = Q/\mathfrak{S}.$$

The proof of the theorem is then completed by choosing $P' = p$ and $Q' = q$ of (4.1.3).

If $P = p$ and $Q = q$, it is seen that (4.2.2) holds. In this case,

$$S(t, s) : f(x) \to e^{is(x + \frac{1}{2}t)}f(x + t)$$

and A of (4.3.5) is given by

$$A : f(x) \to (4\pi)^{\frac{1}{2}} \int_{-\infty}^{\infty} e^{-\frac{1}{2}(x^2 + u^2)}f(u)\,du. \tag{4.3.12}$$

The solutions of $Af = 2\pi f$ form the one-dimensional subspace generated by $f = e^{-\frac{1}{2}x^2}$. In this (irreducible) case only one term appears in the direct sum of (4.3.7) and $\phi_\alpha = \phi = \pi^{-\frac{1}{4}} e^{-\frac{1}{2}x^2}$ and $f_{ts}(x) = \pi^{-\frac{1}{4}} e^{-\frac{1}{2}(x+t)^2 + is(x+\frac{1}{2}t)}$.

As von Neumann points out ([2], p. 578) his proof of Theorem 4.2.1 is related to the treatment of groups given by Frobenius [1] and Peter and Weyl [1].

§ 4.4 The equation $AA^* = A^*A + I$

For definitions, see, e.g., Sz.-Nagy [1]. An operator T is said to be closed if whenever $f_n \in \mathfrak{D}_T$ and both $f_n \to f$ and $Tf_n \to g$ hold, then $f \in \mathfrak{D}_T$ and $g = Tf$. In case T^{**} exists (that is, if T^* is densely defined), then $T^\sim = T^{**}$ is the least closed linear extension of T. In case T is symmetric then $T^\sim = T^{**}$ is closed and symmetric. An operator T is said to be essentially self-adjoint if its closure $T^\sim$ is self-adjoint. If Γ is a set then T/Γ will denote the restriction of T to Γ.

Lemma 4.4.1. *Let A be a closed, linear, densely defined operator on a Hilbert space and suppose that*

$$AA^* = A^*A + I , \qquad (4.4.1)$$

that is, $\mathfrak{D}_{AA^*} = \mathfrak{D}_{A^*A}$ and (4.4.1) holds on this set. Then the operator A^*A is self-adjoint and has a purely discrete spectrum with eigenvalues $\{0, 1, 2, \ldots\}$, each of the same multiplicity.

Proof. Since A is linear, densely defined and closed, the operator $T = A^*A$ is self-adjoint (see Sz.-Nagy [1], p. 30); let its spectral resolution be given by

$$T = A^*A = \int \lambda \, dE_\lambda . \qquad (4.4.2)$$

Since T is non-negative and, by (4.4.1), cannot be the zero operator, there exists some $\mu \in \mathrm{sp}(T)$, $\mu > 0$. Then choose elements $f_n = E(\Delta_n)f_n$ of unit length where $\Delta_n = (\mu - 1/n, \mu + 1/n)$, so that

$$(T - \mu I)f_n = g_n \to 0 . \qquad (4.4.3)$$

Clearly $g_n = E(\Delta_n)g_n \in \mathfrak{D}_T$ and $\|Af_n\|^2 = (f_n, Tf_n) \to \mu$, also $\|Ag_n\|^2 = (g_n, Tg_n) \to 0$. Since $\mathfrak{D}_T \subset \mathfrak{D}_A$ then $AA^*Af_n - \mu Af_n = Ag_n \to 0$ and so, by (4.4.1), $(T - (\mu - 1))Af_n \to 0$. Since $\|Af_n\|^2 \to \mu > 0$, then $\mu - 1 \in \mathrm{sp}(T)$. Thus whenever $\mu \in \mathrm{sp}(T)$ and $\mu > 0$ then $\mu - 1 \in \mathrm{sp}(T)$. Since $T \geq 0$, and hence $\mathrm{sp}(T) \geq 0$, it follows that μ must be a positive integer. Moreover, this argument also implies that 0 is in the point spectrum of T and hence $A^*Af = 0$ holds for some $f \neq 0$. Hence, by (4.4.1), $AA^*f = f$, hence $A^*AA^*f = A^*f$ or $TA^*f = A^*f$. Since $f \neq 0$, then also $A^*f \neq 0$ and so $1 \in \mathrm{sp}(T)$. Similarly, it follows that $\mathrm{sp}(A^*A) = \{0, 1, 2, \ldots\}$.

Let

$$\mathfrak{M}_j = \{f : A^*Af = jf\} . \qquad (4.4.4)$$

It is clear that $\mathfrak{M}_j = \{f : AA^* = (j+1)f\}$ and that

$$A : \mathfrak{M}_{j+1} \to \mathfrak{M}_j, \quad A^* : \mathfrak{M}_j \to \mathfrak{M}_{j+1} \tag{4.4.5}$$

are one-to-one "onto," for each $j = 0, 1, 2, \ldots$. In particular, each $\mathfrak{M}_j$ has the same dimension as $\mathfrak{M}_0$ and the proof of the lemma is complete.

See Putnam [7] and Tillmann [1]. The material of this and the next three sections is based largely on Tillmann's paper [1].

Lemma 4.4.2. Let $\{\phi_{\alpha 0}\}$ be an orthonormal basis for $\mathfrak{M}_0 = \{f : A^*Af = 0\}$. Then the vectors

$$\phi_{\alpha k} = (k!)^{-\frac{1}{2}} A^{*k} \phi_{\alpha 0} \quad (k = 1, 2, \ldots) \tag{4.4.6}$$

exist and, for each $k = 0, 1, 2, \ldots$, $\phi_{\alpha k}$ is an eigenvector of A^*A belonging to the eigenvalue k. Also, the collection $\{\phi_{\alpha k}\}$ is an orthonormal basis for $\mathfrak{H}$.

Proof. The existence of the vectors of (4.4.6) and the relations $A^*A\phi_{\alpha k} = k\phi_{\alpha k}$ are clear from (4.4.1). It is also clear that the spaces $\mathfrak{M}_k$ of (4.4.4) are pairwise orthogonal and that $\mathfrak{H} = \Sigma \oplus \mathfrak{M}_k$. Furthermore, the second relation of (4.4.5) shows that $\{\phi_{\alpha k}\}$ for each fixed k is complete in $\mathfrak{M}_k$, that is, c.l.m. $\{\phi_{\alpha k}\} = \mathfrak{M}_k$. There remains then to be shown that for a fixed $k \geq 1$, $(\phi_{\alpha k}, \phi_{\beta k}) = 1$ or 0 according as $\alpha = \beta$ or $\alpha \neq \beta$. But

$$
\begin{aligned}
(\phi_{\alpha k}, \phi_{\beta k}) &= (k!)^{-1}(A^{*k}\phi_{\alpha 0}, A^{*k}\phi_{\beta 0}) \\
&= ((k-1)!)^{-1}(k^{-1}AA^*A^{*k-1}\phi_{\alpha 0}, A^{*k-1}\phi_{\beta 0}) \\
&= (k^{-1}AA^*\phi_{\alpha, k-1}, \phi_{\beta, k-1}) \\
&= (\phi_{\alpha, k-1}, \phi_{\beta, k-1}) = \ldots = (\phi_{\alpha 0}, \phi_{\beta 0}),
\end{aligned}
$$

and the proof is complete.

Lemma 4.4.3. The operators A and A^* are determined by

$$A\phi_{\alpha k} = k^{\frac{1}{2}}\phi_{\alpha, k-1}, \quad A^*\phi_{\alpha k} = (k+1)^{\frac{1}{2}}\phi_{\alpha, k+1} \quad (\phi_{\alpha, -1} = 0) \tag{4.4.7}$$

for $k = 0, 1, 2, \ldots$, and $\mathfrak{D}_A = \mathfrak{D}_{A^*} = \mathfrak{D}$, where

$$\mathfrak{D} = \left\{ f = \Sigma c_{\alpha k}\phi_{\alpha k} : \sum_{\alpha, k}(k+1)|c_{\alpha k}|^2 < \infty \right\}. \tag{4.4.8}$$

Proof. Relation (4.4.7) follows from the definition of the $\phi_{\alpha k}$. Since A and A^* are closed, it is clear that $\mathfrak{D} \subset (\mathfrak{D}_A \cap \mathfrak{D}_{A^*})$. Suppose that

$$f = \Sigma b_{\alpha k}\phi_{\alpha k} \in \mathfrak{D}_A \quad \text{and} \quad g = \Sigma c_{\alpha k}\phi_{\alpha k} \in \mathfrak{D}_{A^*}$$

and that

$$Af = \Sigma d_{\alpha k}\phi_{\alpha k}, \quad A^*f = \Sigma e_{\alpha k}\phi_{\alpha k}.$$

Then

$$d_{\alpha k} = (Af, \phi_{\alpha k}) = (f, A^*\phi_{\alpha k}) = (k+1)^{\frac{1}{2}}b_{\alpha, k+1}$$

and
$$e_{\alpha k} = (A^* g, \phi_{\alpha k}) = (g, A\phi_{\alpha k}) = k^{\frac{1}{2}} c_{\alpha, k-1}.$$
Hence

$$\sum_{\alpha, k} (k+1)|b_{\alpha k}|^2 = \sum_{\alpha, k}|b_{\alpha k}|^2 + \sum_{\alpha, k}|d_{\alpha, k-1}|^2 < \infty$$

and

$$\sum_{\alpha, k} (k+1)|c_{\alpha k}|^2 = \sum_{\alpha, k}|e_{\alpha, k+1}|^2 < \infty,$$

and so $f, g \in \mathfrak{D}$. This completes the proof of the lemma.

The above results can now be summarized as follows.

Theorem 4.4.1. *Let A be a closed, densely defined, linear transformation satisfying (4.4.1) on a Hilbert space $\mathfrak{H}$. Then $\mathfrak{H}$ can be written as the direct sum $\mathfrak{H} = \Sigma \oplus \mathfrak{H}_\alpha$ of spaces $\mathfrak{H}_\alpha$ such that for each α, $\mathfrak{H}_\alpha$ is an irreducible reducing subspace of A and A^*, of dimension $\aleph_0$, and having a complete orthonormal basis $\{\phi_{\alpha k}\}$, $k = 0, 1, 2, \ldots$, with the property that A and A^* satisfy (4.4.7) and $\mathfrak{D}_A = \mathfrak{D}_{A^*} = \mathfrak{D}$, where $\mathfrak{D}$ is given by (4.4.8).*

§ 4.5 The operators P and Q

Theorem 4.5.1. *Let P, Q be closed, symmetric operators on a Hilbert space $\mathfrak{H}$. Let A be defined by*

$$A = 2^{-\frac{1}{2}}(Q + iP), \tag{4.5.1}$$

and suppose that

$$A \text{ is closed} \tag{4.5.2}$$

and satisfies

$$AA^* = A^*A + I. \tag{4.5.3}$$

Then $A^* = 2^{-\frac{1}{2}}(Q - iP)$, and Q and P satisfy

$$\begin{aligned} Q &= 2^{-\frac{1}{2}}(A^* + A)^\sim \text{ is self-adjoint}, \\ P &= 2^{-\frac{1}{2}}i(A^* - A)^\sim \text{ is self-adjoint}. \end{aligned} \tag{4.5.4}$$

Moreover, if the $\mathfrak{H}_\alpha$ are defined as in Theorem 4.4.1, then each $\mathfrak{H}_\alpha$ reduces P, Q, A, A^* and, if for each α, $P_\alpha = P/\mathfrak{H}_\alpha$ and $Q_\alpha = Q/\mathfrak{H}_\alpha$, then (P_α, Q_α) is an irreducible pair on $\mathfrak{H}_\alpha$. Moreover, the pair (P_α, Q_α) on the space $\mathfrak{H}_\alpha$ is unitarily equivalent to the Heisenberg pair defined by (4.1.6) on the sequential space l^2. In particular, (P_α, Q_α) is an irreducible Schrödinger couple on $\mathfrak{H}_\alpha$.

Proof. One has $A^* \supset 2^{-\frac{1}{2}}(Q^* - iP^*) \supset 2^{-\frac{1}{2}}(Q - iP) \equiv A'$. Since $\mathfrak{D}_A = \mathfrak{D}_{A'} = \mathfrak{D}_P \cap \mathfrak{D}_Q$ and since $\mathfrak{D}_A = \mathfrak{D}_{A^*}$, it follows that $A' = A^*$. According to Theorem 4.4.1, $\mathfrak{D}_A = \mathfrak{D}_{A^*} = \mathfrak{D}$, where $\mathfrak{D}$ is given by (4.4.8). Hence, by (4.5.1), $Qf = 2^{-\frac{1}{2}}(A^* + A)f$ and $Pf = 2^{-\frac{1}{2}}i(A^* - A)f$ for $f \in \mathfrak{D}$. Since Q and P are closed, it follows that $Q \supset 2^{-\frac{1}{2}}(A^* + A)^\sim$ and $P \supset 2^{-\frac{1}{2}}i(A^* - A)^\sim$.

If J is defined on $\mathfrak{H}$ by $J(\Sigma c_{\alpha k} \phi_{\alpha k}) = \Sigma \bar{c}_{\alpha k} \phi_{\alpha k}$, it is seen that J is a conjugation and that the symmetric operator

$$C = (A^* + A)^\sim \tag{4.5.5}$$

is real with respect to J (that is, $f \in \mathfrak{D}_C \Rightarrow Jf \in \mathfrak{D}_C$ and $CJf = JCf$). Consequently (cf. Sz.-Nagy [1], p. 41), C has equal deficiency indices. It will be shown that the domain $\mathfrak{D}_V$ of its Cayley transform $V = V_C$, where $V = (C - iI)(C + iI)^{-1}$, is the entire Hilbert space $\mathfrak{H}$, and hence C is self-adjoint.

In order to prove this, suppose that $\phi_0 \perp \mathfrak{D}_V$. Then, in particular, $(\phi_0, (A^* + A + iI)\phi_{\alpha k}) = 0$. If $\phi_0 = \Sigma d_{\beta j} i^j \phi_{\beta j}$ then, since

$$(A^* + A + iI)\phi_{\alpha k} = (k+1)^{\frac{1}{2}}\phi_{\alpha, k+1} + k^{\frac{1}{2}}\phi_{\alpha, k-1} + i\phi_{\alpha k},$$

it follows that

$$d_{\alpha, k+1} = (k+1)^{-\frac{1}{2}}(d_{\alpha k} + k^{\frac{1}{2}}d_{\alpha, k-1}), \tag{4.5.6}$$

For a fixed $\alpha = \alpha_0$, suppose, if possible, that $d_{\alpha_0, 0} \neq 0$, so that it can be supposed that $d_{\alpha_0, 0} = 1$, hence also $d_{\alpha_0, 1} = 1$. An induction argument then implies that

$$d_{\alpha_0, k+1} \geqq (1 + k^{\frac{1}{2}})(k+1)^{-\frac{1}{2}} \geqq 1$$

and hence $\Sigma |d_{\alpha_0 k}|^2 = \infty$, a contradiction. Hence $d_{\alpha 0} = 0$ for each α and hence, by (4.5.6), $d_{\alpha k} = 0$ for all α, k. Thus $\phi_0 = 0$, hence $\mathfrak{D}_V = \mathfrak{H}$, and so C of (4.5.5) is self-adjoint.

Since $Q \supset 2^{-\frac{1}{2}}C$ and since Q is closed and symmetric then $Q = 2^{-\frac{1}{2}}C$, that is, the first relation of (4.5.4) holds. If

$$\psi_{\alpha k} = i^k \phi_{\alpha k} \text{ then } (A^* + A)\phi_{\alpha k} = (k+1)^{\frac{1}{2}}\phi_{\alpha, k+1} + k^{\frac{1}{2}}\phi_{\alpha, k-1} = 2^{\frac{1}{2}}Q\phi_{\alpha k}$$

and

$$i(A^* - A)\psi_{\alpha k} = (k+1)^{\frac{1}{2}}\psi_{\alpha, k+1} + k^{\frac{1}{2}}\psi_{\alpha, k-1} = 2^{\frac{1}{2}}P\psi_{\alpha k},$$

and hence $(A^* + A)^\sim$ and $i(A^* - A)^\sim$ are unitarily equivalent, hence $i(A^* - A)^\sim$ is self-adjoint, and the second relation of (4.5.4) must hold.

Clearly, each $\mathfrak{H}_\alpha$ reduces P, Q, A, A^*, and the operators (P_α, Q_α) with respect to the orthonormal basis $\{\phi_{\alpha k}\}$ are precisely the Heisenberg matrices (see (4.1.6)) and hence are determined to within unitary equivalence. Furthermore, the pair (P_α, Q_α) is irreducible. For, if $\mathfrak{M} \subset \mathfrak{H}_\alpha$ and if $\mathfrak{M}$ is invariant under P and Q, then $\mathfrak{M}$ is also invariant under A, A^* and $A^* A$. If $E_\mathfrak{M}$ is the projection of $\mathfrak{H}$ onto $\mathfrak{M}$ and if E_j is the projection of $\mathfrak{H}$ onto $\mathfrak{M}_j$ of (4.4.4), then $E_\mathfrak{M}$ commutes with E_j. But $\mathfrak{M}_j \cap \mathfrak{H}_\alpha$ is spanned by the single element $\phi_{\alpha j}$ and hence is one-dimensional. But if $\mathfrak{M} \neq 0$, $\mathfrak{H} = \Sigma \oplus \mathfrak{M}_k$ implies that $\mathfrak{M}_k \cap \mathfrak{M} \neq 0$ for some k and hence $\mathfrak{M}$ contains $\phi_{\alpha k}$. Since $\mathfrak{M}$ is invariant under A and A^*, it is clear (cf. (4.4.7)) that $\mathfrak{M}$ must contain $\phi_{\alpha j}$ for all j, that is, $\mathfrak{M} = \mathfrak{H}_\alpha$.

This completes the proof of Theorem 4.5.1.

§ 4.6 Results of Rellich and Dixmier

The result of von Neumann in § 4.3 yields necessary and sufficient conditions for a pair of self-adjoint operators to be a Schrödinger couple (or a direct sum of such couples) in terms of the Weyl commutation relations (4.2.1). However, in view of the formal equivalence only of (4.1.2) and (4.2.1), the problem for some time remained open in terms of the commutation relation (4.1.2). In this connection, the following result, due to Rellich [2] and Dixmier [1], holds.

Theorem 4.6.1. *On a Hilbert space $\mathfrak{H}$, suppose that*

$$P, Q \text{ are closed and symmetric and } \quad \mathfrak{D}_P \cap \mathfrak{D}_Q \text{ is dense} . \quad (4.6.1)$$

Suppose that there exists a linear set Ω invariant under P and Q, for which

$$PQ - QP = -iI \text{ on } \Omega, \quad \Omega \subset \mathfrak{D}_P \cap \mathfrak{D}_Q, \ \Omega \text{ dense}, \quad (4.6.2)$$

and such that

$$(P^2 + Q^2)/\Omega \text{ is essentially self-adjoint} . \quad (4.6.3)$$

Then P and Q are self-adjoint and (P, Q) is a Schrödinger couple or the direct sum of such couples.

The above formulation is due to Dixmier [1]. Rellich [2] supposed in addition that $M \equiv (P/\Omega)^2 + (Q/\Omega)^2$ was "decomposable," that is, that there exists a spectral family $\{E_\lambda\}$ such that $E(\varDelta)\mathfrak{H} \subset \Omega$ for arbitrary intervals $\varDelta$ and $Mf = \int \lambda dE_\lambda f$ for all $f \in \Omega$.

It can be noted that any Schrödinger pair (P, Q) does satisfy the conditions of the theorem. Indeed, the Schrödinger operators $P = p$ and $Q = q$ of (4.1.3) are seen to satisfy (4.6.2) on $\Omega = \mathfrak{D}_{PQ} \cap \mathfrak{D}_{QP}$. Also the restrictions of p and q to Ω are essentially self-adjoint and, in addition, there exists a dense linear subset Ω_0 of Ω which is invariant under p, q and is such that the restrictions of p, q and $p^2 + q^2$ to Ω_0 are all essentially self-adjoint. For instance, Ω_0 can be chosen to be the set of infinitely differentiable functions $f(x)$ with the property that

$$\lim_{|x| \to \infty} x^n f^{(m)}(x) = 0 \quad \text{for} \quad m, n = 0, 1, 2, \dots .$$

See Foiaş, Gehér and Sz.-Nagy [1].

Proof. Let $Q_0 = Q/\Omega$ and $P_0 = P/\Omega$ and put

$$A_0 = 2^{-\frac{1}{2}}(Q_0 + iP_0), \quad A_0' = 2^{-\frac{1}{2}}(Q_0 - iP_0) . \quad (4.6.4)$$

Then $\mathfrak{D}_{A_0} = \mathfrak{D}_{A_0'} = \Omega$ and clearly $(A_0 f, g) = (f, A_0' g)$ for $f, g \in \Omega$. Hence $A_0' \subset A_0^*$. Since Ω is dense, A_0^{**} exists. Define A by

$$A = A_0^{\widetilde{\ }} \quad (= A_0^{**}), \quad (4.6.5)$$

so that $A \supset A_0$ and $A^* = (A_0^{**})^* = A_0^* \supset A_0'$. For $f \in \Omega$,

$$A_0' A_0 f = \tfrac{1}{2}(P_0^2 f + Q_0^2 f - f) \quad \text{and} \quad A_0 A_0' f = \tfrac{1}{2}(P_0^2 f + Q_0^2 f + f),$$

so that, by (4.6.3), $A_0' A_0$ and $A_0 A_0'$ are essentially self-adjoint. Since A is closed, then $A^* A$ is self-adjoint and clearly $A^* A \supset A_0' A_0$. Hence $A^* A$ is the least closed linear extension of $A_0' A_0$. If $f \in \mathfrak{D}_{A^* A}$, there exist $f_n \in \Omega$ such that $f_n \to f$ and $A^* A f_n (= A_0' A_0 f_n) \to A^* A f$. Also,

$$A A^* f_n = A_0 A_0' f_n = A_0' A_0 f_n + f_n \to A^* A f + f$$

and, since $A A^*$ is self-adjoint (hence closed),

$$f \in \mathfrak{D}_{A A^*} \quad \text{and} \quad A A^* f = A^* A f + f.$$

Consequently, $\mathfrak{D}_{A A^*} \supset \mathfrak{D}_{A^* A}$ and, since $A A^*$ and $A^* A$ are self-adjoint, it is clear that $\mathfrak{D}_{A A^*} = \mathfrak{D}_{A^* A}$ and

$$A A^* = A^* A + I . \tag{4.6.6}$$

It follows from (4.6.6) that $\mathfrak{D}_A = \mathfrak{D}_{A^*} \equiv \mathfrak{D}$ (cf. Sz.-Nagy [1], pp. 30 ff.). Actually even more follows from Theorem 4.5.1 above. For $f \in \mathfrak{D}$, there exist $f_n \in \Omega$ such that $f_n \to f$, $A_0 f_n \to A f$ (by the definition (4.6.5) of A). Also,

$$\begin{aligned}
\| A_0(f_m - f_n) \|^2 &= (A_0' A_0(f_m - f_n), f_m - f_n) \\
&= (\tfrac{1}{2}(P^2 + Q^2 - I)(f_m - f_n), f_m - f_n) \\
&= \tfrac{1}{2}(\| P(f_m - f_n) \|^2 + \| Q(f_m - f_n) \|^2 + \| f_m - f_n \|^2) .
\end{aligned}$$

Hence $P f_n$, $Q f_n$ tend to limits and, since P and Q are closed, $f \in \mathfrak{D}_P \cap \mathfrak{D}_Q$ (and $P f_n \to P f$, $Q f_n \to Q f$). Thus $\mathfrak{D} \subset \mathfrak{D}_P \cap \mathfrak{D}_Q$. Also

$$A_0 f_n = 2^{-\frac{1}{2}}(Q f_n + i P f_n) \to 2^{-\frac{1}{2}}(Q f + i P f)$$

and so

$$A \subset 2^{-\frac{1}{2}}(Q + i P) .$$

Also

$$A^* f_n = A_0' f_n = 2^{-\frac{1}{2}}(Q f_n - i P f_n) \to 2^{-\frac{1}{2}}(Q f - i P f),$$

and since $A^* \supset A_0'$ and A^* is closed, then $A^* \subset 2^{-\frac{1}{2}}(Q - i P)$. Consequently,

$$A = A^{**} \supset [2^{-\frac{1}{2}}(Q - i P)]^* \supset 2^{-\frac{1}{2}}(Q^* + i P^*) \supset 2^{-\frac{1}{2}}(Q + i P) \supset A$$

and so

$$A = 2^{-\frac{1}{2}}(Q + i P) . \tag{4.6.7}$$

The assertion of Theorem 4.6.1 now follows from Theorem 4.5.1 and the proof is complete.

§ 4.7 Results of Tillmann

The following was proved by Tillmann [1].

Theorem 4.7.1. *On a Hilbert space $\mathfrak{H}$ let P and Q satisfy (4.6.1), the relation*

$$(Qf, Pg) - (Pf, Qg) = -i(f, g) \quad \text{for} \quad f, g \in \mathfrak{D}_P \cap \mathfrak{D}_Q, \qquad (4.7.1)$$

and

$$(Q + iP)^* = (Q - iP)^\sim . \qquad (4.7.2)$$

Then P and Q are self-adjoint and (P, Q) is a Schrödinger couple or the direct sum of such couples.

Proof. Note that if A and A' are defined by

$$A = 2^{-\frac{1}{2}}(Q + iP), \quad A' = 2^{-\frac{1}{2}}(Q - iP) \qquad (4.7.3)$$

then clearly

$$A^* = 2^{-\frac{1}{2}}(Q + iP)^* \supset 2^{-\frac{1}{2}}(Q^* - iP^*) = 2^{-\frac{1}{2}}(Q - iP) = A' ,$$

so that $A^* \supset A'$. Relation (4.7.2) then becomes $A^* = (A^*/\mathfrak{D}_P \cap \mathfrak{D}_Q)^\sim$.

First it will be shown that (4.6.1) and (4.7.1) imply (without assuming (4.7.2)) that A and A' of (4.7.3) are closed and satisfy

$$A^* A + \tfrac{1}{2} I = A'^* A' - \tfrac{1}{2} I \supset \tfrac{1}{2}(Q^2 + P^2) . \qquad (4.7.4)$$

To see this, let $f, g \in \mathfrak{D}_A = \mathfrak{D}_{A'} = \mathfrak{D}_P \cap \mathfrak{D}_Q$. Then

$$\begin{aligned}
(Af, Ag) &= \tfrac{1}{2}[(Qf, Qg) + (Pf, Pg) - i((Qf, Pg) - (Pf, Qg))] \\
&= \tfrac{1}{2}[(Qf, Qg) + (Pf, Pg) - (f, g)] .
\end{aligned}$$

Similarly,

$$(A'f, A'g) = \tfrac{1}{2}[(Qf, Qg) + (Pf, Pg) + (f, g)].$$

Hence

$$(A'f, A'g) = (Af, Ag) + (f, g), \quad \|Af\|^2 = \tfrac{1}{2}[\|Qf\|^2 + \|Pf\|^2 - \|f\|^2]$$

and

$$\|A'f\|^2 = \tfrac{1}{2}[\|Qf\|^2 + \|Pf\|^2 + \|f\|^2] .$$

Consequently, since P and Q are closed, also A and A' are closed. Moreover, it also follows that $A'^* A'$ and $A^* A$ have the same domain and that (4.7.4) holds.

Now, assuming (4.7.2), one has $A^* = A'^\sim$, that is, since A' is closed, $A^* = A'$. Hence $AA^* = A^* A + I$ and Theorem 4.5.1 can now be applied to complete the proof of Theorem 4.7.1.

That the assertion of Theorem 4.7.1 can be false if $\mathfrak{D}_{QP} \cap \mathfrak{D}_{PQ}$ is assumed to be dense and if the equation of (4.7.1) is required only for f, g in the domain of $QP - PQ$, that is, for $f, g \in \mathfrak{D}_{QP} \cap \mathfrak{D}_{PQ}$, can be seen as follows. Let $\mathfrak{H}=L^2(0,1)$, $Qf=xf$ (hence $\mathfrak{D}_Q=\mathfrak{H}$), $Pf=-if'$, where $\mathfrak{D}_P=\{f:f,f'\in L^2(0,1), f \text{ absolutely continuous}, f(0)=f(1)\}$. Then P and Q are self-adjoint (see von Neumann [7], p. 137, Stone [1], p. 428). Also (4.6.1) and (4.7.2) hold, while the equation of (4.7.1) is valid for f, g in the (dense) set $\mathfrak{D}_{QP}\cap\mathfrak{D}_{PQ}$. But (P, Q) is not a Schrödinger couple or a direct sum of such couples. Cf. Tillmann [1], p. 261.

§ 4.8 Results of Foiaş, Gehér and Sz.-Nagy

It was noted above that von Neumann's result gives necessary and sufficient conditions in order that a pair of self-adjoint operators form a Schrödinger couple in terms of the Weyl commutation relations (4.2.1). On the other hand the Rellich-Dixmier and Tillmann results deal directly with the original commutation relations (4.1.2) but do not involve (4.2.1). The work of Foiaş, Gehér and Sz.-Nagy [1] considers the connection between (4.1.2) and (4.2.1).

$\{W_t\}$ is said to be a one-parameter (strongly continuous) semi-group of contraction operators if $W_t W_s = W_{t+s}$ for $t, s \geqq 0$, $W_0=I$, $\|W_t\|\leqq 1$, and s-lim $W_t = I$. It is known that the infinitesimal generator $A = s\text{-lim}_{t\to+0}$ $(W_t-I)/t$ exists, is closed, is densely defined, and that $(A-I)^{-1}$ and the infinitesimal cogenerator $W=(A+I)(A-I)^{-1}$ are bounded. Also W is a contraction operator ($\|W\|\leqq 1$) and 1 is not in the point spectrum of W. Conversely, if W denotes any contraction operator not having 1 in its point spectrum, it ·is the infinitesimal cogenerator of a unique contraction semi-group $\{W_t\}$. See Sz.-Nagy and Foiaş [1].

The following is due to Foiaş, Gehér and Sz.-Nagy [1].

Theorem 4.8.1. *Let $\{S_s\}$ and $\{T_t\}$ for $0\leqq s, t<\infty$ be two strongly continuous semi-groups of contraction operators on a Hilbert space $\mathfrak{H}$ with infinitesimal generators A, B defined by the strong limits*

$$A = \text{s-lim}_{s\to+0} (S_s-I)/s, \quad B = \text{s-lim}_{t\to+0} (T_t-I)/t. \qquad (4.8.1)$$

If the relation

$$T_t S_s = e^{its} S_s T_t \qquad (s, t\geqq 0) \qquad (4.8.2)$$

holds then necessarily $\mathfrak{D}_{AB-BA}$ is dense in $\mathfrak{H}$,

$$\mathfrak{D}_{AB-BA} \text{ is invariant under } (A-I)^{-1} \text{ and } (B-I)^{-1}, \qquad (4.8.3)$$

and

$$AB-BA = -iI \qquad (4.8.4)$$

holds on $\mathfrak{D}_{AB-BA}$. Conversely, relation (4.8.2) holds whenever (4.8.4) holds on a linear subset Ω of $\mathfrak{D}_{AB-BA}$ for which either $(B-I)(A-I)\Omega$ or $(A-I)(B-I)\Omega$ is dense in $\mathfrak{H}$.

It is to be noted that (4.8.3) implies that

$$\mathfrak{D}_{AB-BA} \subset (A-I)\mathfrak{D}_{AB-BA}$$

and

$$\mathfrak{D}_{AB-BA} \subset (B-I)\mathfrak{D}_{AB-BA} \subset (B-I)(A-I)\mathfrak{D}_{AB-BA},$$

so that $(B-I)(A-I)\mathfrak{D}_{AB-BA}$ is dense in $\mathfrak{H}$. Similarly, $(A-I)(B-I)\mathfrak{D}_{AB-BA}$ is dense in $\mathfrak{H}$.

The above theorem will be proved in § 4.9 below as a consequence of a more general formulation (Theorem 4.9.1) due to Kato [6].

Theorem 4.8.2. *Let* (P, Q) *be a pair of self-adjoint operators on a Hilbert space* $\mathfrak{H}$. *Suppose that there exists a linear set* $\Omega \subset \mathfrak{D}_{PQ-QP}$ *with the property that* (4.1.2) *holds and*

$$(P+iI)(Q+iI)\Omega \text{ or } (Q+iI)(P+iI)\Omega \text{ is dense in } \mathfrak{H}. \qquad (4.8.5)$$

Then (P, Q) *is a Schrödinger couple or the direct sum of such couples.*

Proof. It is seen from Theorem 4.8.1 that the conditions of Theorem 4.8.2 are necessary and sufficient in order that the unitary operators $U_t = e^{itP}$, $V_s = e^{isQ}$ satisfy the Weyl relations (4.2.1) for s, $t \geq 0$ and, since $U_{-t} = U_t^{-1}$, $V_{-s} = V_s^{-1}$, for all real s, t. The assertion of Theorem 4.8.2 then follows from von Neumann's theorem of § 4.3.

Remark. That the Schrödinger pair (p, q) of (4.1.3), and hence any Schrödinger couple or direct sum of such couples, satisfies the conditions of Theorem 4.8.2 is easily verified.

Corollary. *Let* (P, Q) *denote a pair of closed, symmetric operators on a Hilbert space* $\mathfrak{H}$ *and let* Ω *denote a linear set contained in* $\mathfrak{D}_{PQ-QP}$ *and dense in* $\mathfrak{H}$. *Suppose that* (4.1.2) *holds on* Ω *and that*

$$(P \pm iI)\Omega \supset \Omega \quad \text{and} \quad (Q \pm iI)\Omega \supset \Omega. \qquad (4.8.6)$$

Then (P, Q) *is a Schrödinger couple or the direct sum of such couples.*

It is clear that if (P, Q) is a Schrödinger couple then the conditions of the Corollary are fulfilled.

Proof. Since Ω is dense, it follows from (4.8.6) that the closure P' of P/Ω has deficiency indices $(0, 0)$ and hence P' is self-adjoint. Since $P' \subset P$ and P is symmetric, then $P = P'$ and P is self-adjoint. Similarly, Q is also self-adjoint. Also, it follows from (4.8.6) that $(P+iI)(Q+iI)\Omega \supset \Omega$ and, since Ω is dense, the assertion follows from Theorem 4.8.2.

Before formulating the next result it will be convenient to recall some definitions (see Sz.-Nagy [6]). Let the Hilbert space $\mathfrak{H}$ be a subspace of an extension Hilbert space $\mathfrak{H}'$ and let P denote the orthogonal projection of $\mathfrak{H}'$ onto $\mathfrak{H}$. If T, T' are bounded operators on $\mathfrak{H}$, $\mathfrak{H}'$ respectively, then T' is said to be the dilation of T, and T the compression of T' (see Halmos [2]) if $Tx = PT'x$ for all $x \in \mathfrak{H}$. If $\mathfrak{H}'_1$ and $\mathfrak{H}'_2$ are two extension spaces of $\mathfrak{H}$ and if $\{A_{1\alpha}\}$, $\{A_{2\alpha}\}$ (α in an index set) are families of bounded operators on $\mathfrak{H}_1'$ and $\mathfrak{H}_2'$ respectively, then the "structures" $\{\mathfrak{H}_1', A_{1\alpha}, \mathfrak{H}\}$ and $\{\mathfrak{H}_2', A_{2\alpha}, \mathfrak{H}\}$ are said to be isomorphic if there exists an isometric mapping of $\mathfrak{H}_1'$ onto $\mathfrak{H}_2'$ leaving the elements of $\mathfrak{H}$ invariant and such that $y_1 \to y_2$ implies $A_{1\alpha}y_1 \to A_{2\alpha}y_2$ $(y_1 \in \mathfrak{H}_1', y_2 \in \mathfrak{H}_2')$ for all α.

The following was proved by Foiaş and Gehér [1].

Theorem 4.8.3. *Let* $\{T_t\}$ *and* $\{S_s\}$, *for* $0 \le s, t < \infty$, *be two strongly continuous contraction semi-groups on a Hilbert space* $\mathfrak{H}$ *and suppose that the operators* $T(t), S(s)$, *where* $T(t) = T_t$ *if* $t \ge 0$ *and* $T(t) = T_{-t}^*$ *if* $t < 0$ *and* $S(s) = S_s$ *if* $s \ge 0$ *and* $S(s) = S_{-s}^*$ *if* $s < 0$, *satisfy the Weyl relations*

$$T(t)\,S(s) = e^{its} S(s)\, T(t) \qquad (-\infty < t, s < \infty). \tag{4.8.7}$$

Then there is an extension Hilbert space $\mathfrak{H}'$ *of* $\mathfrak{H}$ *in which there exist two one-parameter strongly continuous groups of unitary operators* $\{U(t)\}$, $\{V(s)\}$, $-\infty < t, s < \infty$, *satisfying the Weyl relations*

$$U(t)\,V(s) = e^{its} V(s)\, U(t) \quad and \quad T(t)\,S(s) = P\,U(t)\,V(s), \quad -\infty < t, s < \infty.$$

Furthermore, $\mathfrak{H}'$ *can be chosen so that it is generated by the elements* $U(t)\,V(s)x$ $(x \in \mathfrak{H}, -\infty < t, s < \infty)$, *in which case the structure* $\{\mathfrak{H}', U(t), V(s), \mathfrak{H}\}$ *is uniquely determined to within isomorphism.*

The proof, which depends upon a result of Sz.-Nagy [3,4] concerning the existence in an extension space of a unitary dilation group associated with a contraction semi-group, will be omitted.

§ 4.9 A result of Kato

Recall that a semi-group $\{T_t\}$, $t \ge 0$, of bounded linear operators on a Banach space B is said to be strongly continuous if T_t is strongly continuous in t and if also $T_0 = I$. If A denotes the infinitesimal generator of T_t then A is densely defined and $T_t = e^{tA}$. The resolvent set of A contains a half-plane $\mathrm{Re}\,(z) > \omega$, where the real number $\omega = \omega_A$ is the type of $\{T_t\}$. (See Hille and Phillips [1], especially Chapters 10, 11. It can be noted that the C_0 convergence defined there on p. 321 is what is presently being supposed. For a justification of the use of the exponential formula $T_t = e^{tA}$, even though A is in general not bounded, see the remarks on p.354.)

The following is due to Kato [6].

Theorem 4.9.1. *Let* $\{e^{sA}\}$ *and* $\{e^{tB}\}$ *be two strongly continuous semi-groups on a Banach space* B *satisfying for some constant* c *the relation*

$$e^{sA} e^{tB} = e^{cst} e^{tB} e^{sA}, \qquad 0 \le s, t < \infty. \tag{4.9.1}$$

Then $\Omega = \mathfrak{D}_{AB} \cap \mathfrak{D}_{BA}$ *is dense in* B *and*

$$(AB - BA)x = cx \quad for \quad x \in \Omega. \tag{4.9.2}$$

Also, $(A-a)(B-b)\Omega = (B-b)(A-a)\Omega = B$ *for all* a, b *satisfying* $\mathrm{Re}(a) > \omega_A$, $\mathrm{Re}(b) > \omega_B$. *Conversely, if there exists a dense linear subset* Ω *of* $\mathfrak{D}_{AB} \cap \mathfrak{D}_{BA}$

for which (4.9.2) holds and either $(A-a)(B-b)\Omega$ or $(B-b)(A-a)\Omega$ is dense in B for some pair a, b satisfying $\mathrm{Re}(a) > \omega_A$ and $\mathrm{Re}(b) > \omega_B$, then (4.9.1) holds.

The theorem reduces to that of Foiaş, Gehér and Sz.-Nagy (Theorem 4.8.1) if B is a Hilbert space $\mathfrak{H}$ and if $\{e^{tA}\}$ and $\{e^{tB}\}$ are contraction semi-groups and $c = -i$.

Proof of first part. Multiplication of (4.9.1) by e^{-as} followed by an integration with respect to s on $(0, \infty)$ yields

$$(A-a)^{-1} e^{tB} = e^{tB}(A+ct-a)^{-1}, \qquad t \geq 0, \qquad (4.9.3)$$

whenever $\mathrm{Re}(a) > \omega_A$ and $\mathrm{Re}(a-ct) > \omega_A$. Differentiation of (4.9.3) with respect to t followed by setting $t=0$ leads to

$$B(A-a)^{-1} \supset (A-a)^{-1}B + c(A-a)^{-2}$$

and hence, for $\mathrm{Re}(a) > \omega_A$ and $\mathrm{Re}(b) > \omega_B$,

$$(A-a)^{-1}(B-b)^{-1} = (B-b)^{-1}(A-a)^{-1} + c(B-b)^{-1}(A-a)^{-2}(B-b)^{-1}.$$
$$(4.9.4)$$

If $y \in B$ and

$$x = (A-a)^{-1}(B-b)^{-1} y, \qquad (4.9.5)$$

then

$$y = (B-b)(A-a)x$$

and hence, by (4.9.4),

$$x = (B-b)^{-1}(A-a)^{-1}(y+cx).$$

Hence

$$x \in \mathfrak{D}_{(A-a)(B-b)} \quad \text{and} \quad (A-a)(B-b)x = (B-b)(A-a)x + cx.$$

So $x \in \mathfrak{D}_{AB} \cap \mathfrak{D}_{BA} \equiv \Omega$ and $(AB - BA)x = cx$. It is clear that any element x of Ω can be expressed in the form (4.9.5) by letting $y = (B-b)(A-a)x$, and so relation (4.9.2) holds. Also, since $y \in B$ is arbitrary, then $(B-b)\cdot(A-a)\Omega = B$ and so $\Omega = (A-a)^{-1}(B-b)^{-1}B$. Since A and B are densely defined, then Ω is dense. In like manner it follows that $(A-a)(B-b)\Omega = B$ and the proof of the first part of Theorem 4.9.1 is complete.

Proof of second part. Let a_0, b_0 denote constants for which $\mathrm{Re}(a_0) > \omega_A$, $\mathrm{Re}(b_0) > \omega_B$ and $(B-b_0)(A-a_0)\Omega$ is dense in B. If $x \in \Omega$ and $y = (B-b_0)\cdot(A-a_0)x$, then by (4.9.2),

$$y = (A-a_0)(B-b_0)x - cx$$

and consequently

$$(A-a_0)^{-1}(B-b_0)^{-1} y = x = (B-b_0)^{-1}(A-a_0)^{-1}(y+cx)$$
$$= (B-b_0)^{-1}(A-a_0)^{-1} y$$
$$+ c(B-b_0)^{-1}(A-a_0)^{-2}(B-b_0)^{-1} y.$$

Since the y's are dense then (4.9.4) holds when $a=a_0$ and $b=b_0$.

Next, it will be shown that

$$(A-a)^{-n}(B-b)^{-1}=(B-b)^{-1}(A-a)^{-n}+nc(B-b)^{-1}(A-a)^{-n-1}(B-b)^{-1}$$
(4.9.6)

holds for $a=a_0$ and $b=b_0$ and $n=1, 2, \ldots$. The result has already been established for $n=1$. The induction from n to $n+1$ then proceeds as follows. If

$$M = (A-a_0)^{-1} \quad \text{and} \quad N = (B-b_0)^{-1}$$

then

$$\begin{aligned}
M^{n+1}N - NM^{n+1} &= M^n(MN-NM)+(M^nN-NM^n)M \\
&= cM^nNM^2N+ncNM^{n+1}NM \\
&= c(NM^n+ncNM^{n+1}N)M^2N \\
&\quad + ncNM^{n+1}(MN-cNM^2N) \\
&= (n+1)cNM^{n+2}N .
\end{aligned}$$

Thus (4.9.6) holds for $a=a_0$ and $b=b_0$ and $n=1, 2, \ldots$.

Since

$$(A-a)^{-1} = \sum_{k=1}^{\infty} (a-a_0)^{k-1}(A-a_0)^{-k}$$

and

$$(A-a)^{-2} = \sum_{k=1}^{\infty} k(a-a_0)^{k-1}(A-a_0)^{-k-1}$$

it follows from (4.9.6) for $a=a_0$ and $b=b_0$ that (4.9.4) holds for $b=b_0$ and $|a-a_0|$ sufficiently small. Since $(A-a)^{-1}$ is analytic for $\mathrm{Re}(a)>\omega_A$ then (4.9.4) must hold for $\mathrm{Re}(a)>\omega_A$ and $b=b_0$. An $(n-1)$-fold differentiation of (4.9.4) (when $b=b_0$) with respect to a then shows that (4.9.6) holds for $b=b_0$ and $\mathrm{Re}(a)>\omega_A$.

If (4.9.6) is multiplied by $(-a)^n$ and if $a=n/s$ $(s>0)$ then

$$(1-n^{-1}sA)^{-n}(B-b)^{-1} = (B-b)^{-1}(1-n^{-1}sA)^{-n}$$
$$- cs(B-b)^{-1}(1-n^{-1}sA)^{-n-1}(B-b)^{-1} \quad \text{for } b=b_0 \text{ and } n>s\omega_A .$$

But

$$\text{s-}\lim_{n\to\infty}(1-n^{-1}sA)^{-n} = e^{sA}$$

(Hille and Phillips [1], p. 362) and so

$$\begin{aligned}
e^{sA}(B-b)^{-1} &= (B-b)^{-1}e^{sA} - cs(B-b)^{-1}e^{sA}(B-b)^{-1} \\
&= (B-b)^{-1}e^{sA}(B-cs-b)(B-b)^{-1} \quad \text{for } b=b_0, s\geq 0 .
\end{aligned}$$

If s is chosen so small that $\mathrm{Re}(b_0+cs)>\omega_B$, it follows that

$$e^{sA}(B-cs-b)^{-1} = (B-b)^{-1}e^{sA} \qquad (4.9.7)$$

for $b=b_0$, and hence also

$$e^{sA}(B-cs-b)^{-n} = (B-b)^{-n}e^{sA} \qquad (n=1, 2, \ldots) \qquad (4.9.8)$$

for $b=b_0$. Using the power series representation for $(B-b)^{-1}$ and $(B-cs-b)^{-1}$ near $b=b_0$ one concludes from (4.9.8) (where $b=b_0$) and the argument used earlier that (4.9.7) holds for $|b-b_0|$ sufficiently small and, by analytic continuation, for all b satisfying $\mathrm{Re}(b) > \omega_B$ and $\mathrm{Re}(b+cs) > \omega_B$. A differentiation of (4.9.7) with respect to b shows that (4.9.8) holds also for such b. If one multiplies both sides of (4.9.8) by $(-b)^n$, lets $b=n/t$ where $t>0$, and then lets $n \to \infty$, one obtains $e^{sA}e^{t(B-cs)} = e^{tB}e^{sA}$, that is (4.9.1), for sufficiently small positive s depending on b_0. The semi-group property of e^{sA} then implies that (4.9.1) holds for all $s \geq 0$ and the proof of Theorem 4.9.1 is complete.

As Kato further notes ([6], p. 275), the relations $(A-a)\Omega = (B-b)^{-1}B = \mathfrak{D}_B$ and $(B-b)\Omega = \mathfrak{D}_A$ in the first part of the proof, where $\Omega = \mathfrak{D}_{AB} \cap \mathfrak{D}_{BA}$, imply that both $(A-a)\Omega$ and $(B-b)\Omega$ are dense in B. An open problem is whether the existence of a dense linear subset Ω of $\mathfrak{D}_{AB} \cap \mathfrak{D}_{BA}$ for which (4.9.2) holds and for which $(A-a)\Omega$ and $(B-b)\Omega$ are dense in B for some pair a, b satisfying $\mathrm{Re}(a) > \omega_A$ and $\mathrm{Re}(b) > \omega_B$, implies the validity of (4.9.1).

§ 4.10 Results of Kristensen, Mejlbo and Poulsen

Kristensen, Mejlbo and Poulsen[1, 2] investigate the commutation relations of quantum mechanics by a different approach, using the theory of topological vector spaces. Their idea roughly is to replace the underlying Hilbert space by a smaller space (but with a larger dual space) on which the "operators" are continuous. Along these lines there will be mentioned only one result (Theorem 4.10.1 below), due to Mejlbo [1], and which generalizes a result in Kristensen, Mejlbo and Poulsen [1].

By the Schwartz space $\mathscr{S}$ will be meant the set of complex-valued functions $\phi = \phi(t)$ defined on $-\infty < t < \infty$ of class C^∞, and such that

$$\sup_t |t^n \phi^{(m)}(t)| < \infty \qquad (m, n = 0, 1, \ldots), \qquad (4.10.1)$$

with topology defined by the semi-norms of (4.10.1). (Cf., e.g., Hörmander [1], p. 18.) Further let there be defined on $\mathscr{S}$ the usual scalar product

$$(\phi, \psi) = \int_{-\infty}^{\infty} \phi(t)\,\bar{\psi}(t)\,dt$$

and the operators p, q satisfying (4.1.3), that is

$$(p\phi)(t) = -i\phi'(t), \quad (q\phi)(t) = t\phi(t). \qquad '(4.10.2)$$

Theorem 4.10.1. *Let W be a non-trivial vector space of vectors x, y, …, over the complex numbers on which there is defined a scalar product* (x, y) *with norm* $\|x\| = (x, x)^{\frac{1}{2}}$. *Let P, Q be linear operators mapping W into itself and satisfying the symmetry conditions* $(Px, y) = (y, Px), (Qx, y) = (x, Qy)$ *for all x, y in W; the commutation relation* $PQ - QP = -iI$; *and the relations* $(P \pm iI)W = (Q \pm iI)W = W$. *Suppose that W is complete in the topology determined by the semi-norms of the form* $\|Ax\|$ *where A belongs to the algebra generated by P and Q. Suppose that the only non-trivial subspace of W closed in this topology and invariant under P, Q,* $(P \pm iI)^{-1}, (Q \pm iI)^{-1}$ *(all operators existing by the hypotheses made) is W itself. Then there exists a one-to-one, bicontinuous, linear inner product preserving mapping J of W onto* $\mathscr{S}$ *such* $JP = pJ$ *and* $JQ = qJ$.

The proof can be found in Mejlbo [1] and will be omitted. As is noted there, if W is embedded in a Hilbert space $\mathfrak{H}$, the theorem yields a result concerning operators P and Q on $\mathfrak{H}$ which is similar to, but weaker than, one of Foiaş, Gehér and Sz.-Nagy [1] (see the Corollary of Theorem 4.8.2).

§ 4.11 Systems with n ($< \infty$) degrees of freedom

The theorems of the preceding sections of von Neumann, Rellich, Dixmier, Tillmann have also been treated for the case of a finite number of degrees of freedom, that is, for the case where $P_1, \ldots, P_n, Q_1, \ldots, Q_n$ satisfy $P_k Q_k - Q_k P_k = -iI$ ($k = 1, \ldots, n$) and operators with different subscripts commute (in some appropriate sense). Also, Foiaş, Gehér and Sz.-Nagy [1] and Mejlbo [1] remark that their results (see §§ 4.8, 4.10) for $n = 1$ generalize to the present case.

A summary of the results for a finite number of degrees of freedom, without proofs, will be given in this section.

A system $\{P_1, \ldots, P_n, Q_1, \ldots, Q_n\}$ of self-adjoint operators on a Hilbert space $\mathfrak{H}$ will be called a Schrödinger n-system if $\mathfrak{H} = \Sigma \oplus \mathfrak{H}_\alpha$ where each $\mathfrak{H}_\alpha$ reduces all P_j, Q_j and the system $(P_1, \ldots, P_n, Q_1, \ldots, Q_n)$ is, in each $\mathfrak{H}_\alpha$, irreducible and unitarily equivalent to the Schrödinger system $(p_1, \ldots, p_n, q_1, \ldots, q_n)$ in the case of n degrees of freedom.

The next result is due to von Neumann [2].

Theorem 4.11.1. *Let* $(P_1, \ldots, P_n, Q_1, \ldots, Q_n)$ *be self-adjoint operators on a Hilbert space* $\mathfrak{H}$ *and put* $U_k(t) = e^{itP_k}$ *and* $V_k(s) = e^{isQ_k}$ *for* $k = 1, \ldots, n$. *If the Weyl relations*

$$U_k(t) U_k(s) = U_k(t+s), \quad V_k(t) V_k(s) = V_k(t+s), \qquad (4.11.1)$$

$$U_j(t) U_k(s) = U_k(s) U_j(t), \quad V_j(t) V_k(s) = V_k(s) V_j(t),$$

$$U_j(t) V_k(s) = e^{i\delta_{jk}ts} V_k(s) U_j(t)$$

hold, then $(P_1, \ldots, P_n, Q_1, \ldots, Q_n)$ *is a Schrödinger n-system.*

The proof of Theorem 4.11.1 is a straightforward generalization of that for the case $n=1$ (see Theorem 4.3.1).

The next result is due to Dixmier [1].

Theorem 4.11.2. *On a Hilbert space $\mathfrak{H}$ suppose that*

$$P_1, \ldots, P_n, Q_1, \ldots, Q_n \text{ are closed and symmetric.} \qquad (4.11.2)$$

Suppose that there exists a dense, linear set Ω contained in

$$\mathfrak{D}_{P_1} \cap \ldots \cap \mathfrak{D}_{P_n} \cap \mathfrak{D}_{Q_1} \cap \ldots \cap \mathfrak{D}_{Q_n},$$

invariant under $P_1, \ldots, P_n, Q_1, \ldots, Q_n$, and such that for $j, k = 1, \ldots, n$,

$$P_j Q_k - Q_k P_j = -iI, \quad P_j P_k - P_k P_j = 0, \qquad (4.11.3)$$
$$Q_j Q_k - Q_k Q_j = 0 \text{ hold on } \Omega,$$

and

$$(P_j^2 + Q_k^2)/\Omega, \quad (P_j^2 + P_k^2)/\Omega, \quad (Q_j^2 + Q_k^2)/\Omega \qquad (4.11.4)$$

are essentially self-adjoint.

Then $(P_1, \ldots, P_n, Q_1, \ldots, Q_n)$ is a Schrödinger n-system.

Rellich [2] proved the assertion of Theorem 4.11.2 when the hypothesis (4.11.4) is replaced by

$$N \equiv \sum_{j=1}^{n} [(P_j/\Omega)^2 + (Q_j/\Omega)^2] \text{ is decomposable,} \qquad (4.11.5)$$

that is, there exists a spectral family $\{E_\lambda\}$ such that for any interval Δ,

$$f \in \mathfrak{D}_N \Rightarrow E(\Delta)f \in \mathfrak{D}_N \quad \text{and} \quad Nf = \int \lambda \, dE_\lambda f. \qquad (4.11.6)$$

It turns out that the condition (4.11.5) implies that each $N_j = (P_j/\Omega)^2 + (Q_j/\Omega)^2$ is essentially self-adjoint. Dixmier [1] proves this as follows. Obviously it is sufficient to consider N_1 only. If $f \in \Omega$, then

$$(Nf, Nf) = \left(N_1 f + \sum_{j=2}^{n} N_j f, \ N_1 f + \sum_{j=2}^{n} N_j f \right)$$

$$= (N_1 f, N_1 f) + \left(\sum_{j=2}^{n} N_j f, \ \sum_{j=2}^{n} N_j f \right)$$

$$+ \left(N_1 f, \ \sum_{j=2}^{n} N_j f \right) + \left(\sum_{j=2}^{n} N_j f, \ N_1 f \right).$$

But

$$\left(N_1 f, \ \sum_{j=2}^{n} N_j f \right) = \left(P_1^2 f + Q_1^2 f, \ \sum_{j=2}^{n} (P_j^2 f + Q_j^2 f) \right)$$

$$= \sum_{j=2}^{n} (\|P_1 P_j f\|^2 + \|P_1 Q_j\|^2 + \|Q_1 P_j f\|^2$$

$$+ \|Q_1 Q_j f\|^2) \geq 0,$$

and hence $\|Nf\| \geqq \|N_1 f\|$. Now $E_\lambda = 0$ for $\lambda < 0$ and $E_\lambda(\mathfrak{H}) \subset \Omega$ for all λ. If $f \in E_\mu(\mathfrak{H})$, then

$$\|N^k N_1 f\| = \|N_1 N^k f\| \leqq \|N^{k+1} f\| \leqq \mu^{k+1} \|f\|,$$

from which it follows that $N_1 f \in E_\mu(\mathfrak{H})$. Hence each subspace $E_\mu(\mathfrak{H})$ reduces N_1, and the restriction of N_1 to this space is a bounded self-adjoint operator. Since $\mathfrak{H} = \Sigma(E_{n+1} - E_n)(\mathfrak{H})$, it follows that N_1 is essentially self-adjoint.

As Dixmier remarks, it does not seem to follow from the hypotheses of Theorem 4.11.2 that, for instance, $(P_1/\Omega)^2 + (P_2/\Omega)^2$ is essentially self-adjoint, so that Rellich's result is not contained in the Dixmier theorem (except when $n=1$). However, Kilpi [2] has proved, using results on the complex moment problem (Kilpi [1]), that the assertion of Dixmier's theorem (Theorem 4.11.2) remains valid if (4.11.4) is replaced by the weaker requirement

$$(P_j^2 + Q_j^2)/\Omega \quad (j = 1, \ldots, n) \text{ is essentially self-adjoint,} \quad (4.11.7)$$

that is, the following result.

Theorem 4.11.3. *Relations* (4.11.2), (4.11.3) *and* (4.11.7) *imply that* $(P_1, \ldots, P_n, Q_1, \ldots, Q_n)$ *is a Schrödinger n-system.*

Rellich's theorem for n degrees of freedom is contained in Theorem 4.11.3.

For any pair of self-adjoint operators T_1 and T_2 with spectral resolutions $T_k = \int \lambda \, dE_{k\lambda}$ let $T_1 \smile T_2$ (cf. Riesz and Sz.-Nagy [1]) signify that $E_{1\lambda} E_{2\mu} = E_{2\mu} E_{1\lambda}$ for all (real) λ and μ.

The following two theorems are due to Tillmann [1,2].

Theorem 4.11.4. *On a Hilbert space $\mathfrak{H}$, suppose that* (4.11.2) *holds, that*

$$(Q_k f, P_k g) - (P_k f, Q_k g) = -i(f, g) \text{ for } f, g \in \mathfrak{D}_{P_k} \cap \mathfrak{D}_{Q_k} \quad (4.11.8)$$

holds for $k = 1, \ldots, n$ *and that*

$$(Q_k + iP_k)^* = (Q_k - iP_k)^\sim (= ([Q_k + iP_k]^*/\mathfrak{D}_{P_k} \cap \mathfrak{D}_{Q_k})^\sim), \quad (4.11.9)$$

so that, in particular, by Theorem 4.7.1, all P_j, Q_j must be self-adjoint. Finally, when $n > 1$, suppose that

$$P_k \smile P_j, Q_k \smile Q_j, P_k \smile Q_j \text{ for } k \neq j. \quad (4.11.10)$$

Then $(P_1, \ldots, P_n, Q_1, \ldots, Q_n)$ *is a Schrödinger n-system.*

Tillmann [2] shows that (4.11.10) is implied by the following conditions, analogous to (4.11.8).

$$(P_k f, Q_j g) - (Q_j f, P_k g) = 0 \text{ for } f, g \in \mathfrak{D}_{P_k} \cap \mathfrak{D}_{Q_j} \quad (j \neq k), \quad (4.11.11)$$
$$(P_k f, P_j g) - (P_j f, P_k g) = 0 \text{ for } f, g \in \mathfrak{D}_{P_k} \cap \mathfrak{D}_{P_j},$$
$$(Q_k f, Q_j g) - (Q_j f, Q_k g) = 0 \text{ for } f, g \in \mathfrak{D}_{Q_k} \cap \mathfrak{D}_{Q_j},$$

and $\qquad (R+iS)^* = (R-iS)^\sim \quad (=([R+iS]^*/\mathfrak{D}_R \cap \mathfrak{D}_S)^\sim),$ $\qquad$ (4.11.12)

where R, S are any two of the operators P_j, Q_j with distinct indices. Thus one has the following result.

Theorem 4.11.5. *Relations* (4.11.2), (4.11.8), (4.11.9) (4.11.11) *and* (4.11.12) *imply that* $(P_1, \ldots, P_n, Q_1, \ldots, Q_n)$ *is a Schrödinger n-system.*

§ 4.12 Anticommutation relations

It will be convenient to recall the notion of a partial isometry U; cf. Murray and von Neumann [1], Kuroda [2]. In a Hilbert space $\mathfrak{H}$, let $\mathfrak{M}$ and $\mathfrak{N}$ be subspaces of the same dimension and let U be a bounded operator. Then U is said to be partially isometric with initial set $\mathfrak{M}$ and final set $\mathfrak{N}$ if $U\mathfrak{M} = \mathfrak{N}$, $\|Ux\| = \|x\|$ for $x \in \mathfrak{M}$ and $Ux = 0$ for $x \in \mathfrak{M}^\perp$, that is, if $U^*U = P$, $UU^* = Q$ are orthogonal projections satisfying $P\mathfrak{H} = \mathfrak{M}$, $Q\mathfrak{H} = \mathfrak{N}$. U^* is also partially isometric with initial set $\mathfrak{N}$ and final set $\mathfrak{M}$. In addition, if $VU = P$ where V is a partially isometric operator with initial set $\mathfrak{N}$ then necessarily $V = U^*$. (If $\mathfrak{M} = \mathfrak{H}$ then $U^*U = I$ and U is isometric; in case also $\mathfrak{N} = \mathfrak{H}$ then U is unitary.)

The material of the remainder of this section as well as that of §§ 4.13, 4.14 is based on the treatment in Tillmann [1].

In the quantization of wave fields for particles satisfying Fermi-Dirac statistics the particles are described through (closed) operators B and B^* on a Hilbert space $\mathfrak{H}$ satisfying the anticommutation relations

$$BB^* + B^*B = I, \quad B^2 = B^{*2} = 0. \qquad (4.12.1)$$

The next result is essentially due to Jordan and Wigner [1] and is a uniqueness theorem for the operators B.

Theorem 4.12.1. *On a Hilbert space $\mathfrak{H}$ let B be a closed operator satisfying* (4.12.1) *and let F^0, F^1 be defined by*

$$F^0 = BB^*, \quad F^1 = B^*B = I - F^0. \qquad (4.12.2)$$

Then F^0, F^1 are orthogonal projections and, if $\mathfrak{H}^0 = F^0\mathfrak{H}$ and $\mathfrak{H}^1 = F^1\mathfrak{H}$ $(= \mathfrak{H}^{0\perp})$, then $B[B^*]$ is a partial isometry with initial set $\mathfrak{H}^1[\mathfrak{H}^0]$ and final set $\mathfrak{H}^0[\mathfrak{H}^1]$. Moreover if $\{\phi_{v0}\}$ is an orthonormal basis for $\mathfrak{H}^0$ then there exists an orthonormal basis $\{\phi_{v1}\}$ for $\mathfrak{H}^1$ such that B, B^* satisfy

$$B^* \phi_{v0} = \phi_{v1}, \quad B^* \phi_{v1} = 0 \qquad (4.12.3)$$

and

$$B\phi_{v0} = 0, \quad B\phi_{v1} = \phi_{v0}. \qquad (4.12.4)$$

Each two-dimensional space $\mathfrak{H}^{(v)}$ spanned by ϕ_{v0} and ϕ_{v1} reduces B and B^*, and hence B, B^* are unitarily equivalent to a direct sum of matrices

$$B^{(v)} = \begin{pmatrix} 0 & 1 \\ 0 & 0 \end{pmatrix}, \quad B^{(v)*} = \begin{pmatrix} 0 & 0 \\ 1 & 0 \end{pmatrix}. \qquad (4.12.5)$$

Proof. It is clear from the first part of (4.12.1) that, since B is closed, B must be bounded. Also the second part of (4.12.1) and (4.12.2) imply that $F^1 F^0 = F^0 F^1 = 0$ and $(F^0)^2 = F^0$, $(F^1)^2 = F^1$, so that F^0 and F^1 are self-adjoint, idempotent, hence orthogonal projections on closed subspaces $\mathfrak{H}^0$ and $\mathfrak{H}^1$ respectively. Clearly $\mathfrak{H}^0 \perp \mathfrak{H}^1$ and $\mathfrak{H} = \mathfrak{H}^0 \oplus \mathfrak{H}^1$. Since $B^* B$ is a projection, B is a partial isometry. Moreover, if $x_1 \in \mathfrak{H}^1$, then $\|Bx_1\|^2 = (B^* Bx_1, x_1) = \|x_1\|^2$ and, since $F^1 Bx_1 = B^*(B^2)x_1 = 0$, also $Bx_1 \in \mathfrak{H}^0$, so that B has initial set $\mathfrak{H}^1$ and final set $\mathfrak{H}^0$, while B^* has initial set $\mathfrak{H}^0$ and final set $\mathfrak{H}^1$. For any orthonormal basis $\{\phi_{v0}\}$ for $\mathfrak{H}^0$ let $\phi_{v1} = B^* \phi_{v0}$. Then (4.12.3) and (4.12.4) hold, so that $\{\phi_{v1}\}$ is an orthonormal basis for $\mathfrak{H}^1$. The last part of the theorem is now clear and the proof is complete.

§ 4.13 General systems

Next there will be considered systems $\mathscr{S}$ of operators satisfying either commutation relations (boson fields) or anticommutation relations (fermion fields). Let

$$\mathscr{S} = \{P_\mu, Q_\mu, B_\lambda, B_\lambda^*\}_{\mu \in M, \, \lambda \in N}, \tag{4.13.1}$$

where M and N are at most denumerable index sets, denote a system of operators on a Hilbert space $\mathfrak{H}$ satisfying the following conditions:

(i) For each $\mu \in M$, (P_μ, Q_μ) satisfies (4.6.1), (4.7.1) and (4.7.2), so that, by Theorem 4.7.1, (P_μ, Q_μ) is a Schrödinger couple.

(ii) For each $\lambda \in N$, B_λ is a bounded operator satisfying

$$B_\lambda B_\kappa^* + B_\kappa^* B_\lambda = \delta_{\kappa\lambda} I, \quad B_\lambda B_\kappa + B_\kappa B_\lambda = 0.$$

(iii) P_μ and Q_μ commute with P_κ and Q_κ for $\mu \neq \kappa$ $(\mu, \kappa \in M)$ and with B_λ, B_λ^* for all $\lambda \in N$; that is, the projections of the spectral families of P_μ and Q_μ commute with those of P_κ and Q_κ $(\mu \neq \kappa)$ and commute in the ordinary sense with the bounded operators B_λ, B_λ^*.

It was shown earlier that

$$F_\lambda^0 = B_\lambda B_\lambda^* = I - B_\lambda^* B_\lambda = I - F_\lambda^1 \tag{4.13.2}$$

are orthogonal projections and that B_λ, B_λ^* are partial isometries. It is easily shown from (ii) that for arbitrary $\lambda, \kappa \in N$, $F_\lambda^i F_\kappa^j = F_\kappa^j F_\lambda^i$ $(i, j = 0 \text{ or } 1)$ and that if $\lambda \neq \kappa$, F_κ^j commutes with B_λ and B_λ^*.

Furthermore, if E_0 denotes any projection of the spectral family of either P_μ or Q_μ then it follows from (iii) that if $\mu \neq v$, $\mathfrak{H}_0 = E_0(\mathfrak{H})$ reduces each of the operators $P_v, Q_v, B_\lambda, B_\lambda^*$. In particular, E_0 commutes with F_λ^j, and corresponding to the decomposition $\mathfrak{H} = \mathfrak{H}_0 \oplus \mathfrak{H}_1$ $(\mathfrak{H}_1 = \mathfrak{H}_0^\perp)$ one has a representation

$$P_v = P_{v0} \oplus P_{v1}, \quad Q_v = Q_{v0} \oplus Q_{v1}. \tag{4.13.3}$$

Similar representations for P_ν^2, Q_ν^2 and $P_\nu^2 + Q_\nu^2$ lead to

$$(P_\nu^2 + Q_\nu^2)^\sim = (P_{\nu 0}^2 + Q_{\nu 0}^2)^\sim \oplus (P_{\nu 1}^2 + Q_{\nu 1}^2)^\sim = A_\nu A_\nu^* - \tfrac{1}{2} I , \quad (4.13.4)$$

and hence the self-adjoint operator $A_\nu A_\nu^*$, where $A_\nu = 2^{-\frac{1}{2}}(Q_\nu + iP_\nu)$ and $A_\nu^* = 2^{-\frac{1}{2}}(Q_\nu - iP_\nu)$, is reduced by $\mathfrak{H}_0$. Hence, for $\mu \neq \nu$, $A_\nu A_\nu^*$ commutes with P_μ, Q_μ, A_μ, A_μ^*, $A_\mu A_\mu^*$. By the results of § 4.4, $A_\mu A_\mu^*$ can be expressed as

$$A_\mu A_\mu^* = \sum_{m=0}^\infty (m+1) E_\mu^m . \quad (4.13.5)$$

Since F_λ^j commutes with P_μ, Q_μ, hence with $A_\mu A_\mu^*$, it follows that F_λ^j commutes with E_μ^m ($j = 0, 1$; $m = 1, 2, \ldots$; $\mu \in M$, $\lambda \in N$). The above results can be summarized as follows.

Theorem 4.13.1. *Let* (i), (ii), (iii) *hold. Then the projections* $\{E_\mu^m, F_\lambda^j\}$ *where* E_μ^m, F_λ^j *are defined by* (4.13.5) *and* (4.13.2) *form a commutative system of operators. Each of the operators* P_ν, Q_ν, A_ν, A_ν^* *is reduced by all* E_μ^m *with* $\mu \neq \nu$ *and by all* F_λ^j. *Each of the operators* B_λ, B_λ^* *is reduced by all* E_μ^m *and by all* F_κ^j *with* $\kappa \neq \lambda$.

The eigenvalues $0, 1, 2, \ldots$ of $A_\mu^* A_\mu$ are known as the occupation numbers for a particle in the state μ. Thus in quantization according to Bose-Einstein statistics an arbitrarily large number of particles can exist in the same state. But $B_\lambda^* B_\lambda$ can have only the eigenvalues 0 and 1, these being the occupation numbers of particles in the state λ. This fact, that not more than one particle obeying Fermi-Dirac statistics can be in a given state, is a manifestation of the Pauli exclusion principle. See, e.g., Bogoliubov and Shirkov [1], p. 112.

§ 4.14 A uniqueness theorem

Theorem 4.14.1. *Let* $\mathscr{S}$ *be a system defined by* (4.13.1) *and satisfying* (i), (ii) *and* (iii). *In addition suppose that*

(iv) *there exists a "vacuum state," that is, there is an element* $\phi_0 \in \mathfrak{H}$, $\|\phi_0\| = 1$, *in the domain of all* A_μ, B_λ ($\mu \in M$, $\lambda \in N$) *for which* $A_\mu \phi_0 = 0$, $B_\lambda \phi_0 = 0$.

Next, suppose that

(v) $\mathscr{S}$ *is an irreducible system.*

Then $\mathscr{S}$ *is uniquely determined to within unitary equivalence* (*and is given by* (4.14.2) *and* (4.14.3) *below*).

Proof. It is clear from $A_\mu A_\mu^* = A_\mu^* A_\mu + I$ and from $B_\lambda B_\lambda^* = I - B_\lambda^* B_\lambda$ that repeated applications of B_λ^* and A_μ^* to ϕ_0 lead to vectors in the set of simultaneous eigenvectors of the family $\{A_\mu A_\mu^*, B_\lambda B_\lambda^*\}$. For each sequence $m = \{m_\mu\} = \{m_{\mu_1}, \ldots, m_{\mu_k}, \ldots\}$ where $m_{\mu_i} \in \{0, 1, 2, \ldots\}$ and $\Sigma m_{\mu_i} < \infty$ and each sequence $n = \{n_\lambda\} = \{n_{\lambda_1}, \ldots, n_{\lambda_j}, \ldots\}$, where

$$n_{\lambda_i} \in \{0, 1\} \text{ and } \Sigma n_{\lambda_i} < \infty ,$$

consider $(m, n) = (m_{\mu_1}, \ldots, m_{\mu_k}, \ldots; n_{\lambda_1}, \ldots, n_{\lambda_j}, \ldots)$. Then, since only a finite number of m_μ's and n_λ's are different from 0, the vector

$$\phi_{m,n} = \prod_\mu (m_\mu!)^{-\frac{1}{2}} (A_\mu^*)^{m_\mu} \prod_\lambda (B_\lambda^*)^{n_\lambda} \phi_0 \qquad (4.14.1)$$

is defined. Suppose in addition that the λ-factors occur in the order of increasing λ. It follows from Lemma 4.4.3 and Theorem 4.12.1 that for arbitrary λ_0 and μ_0

$$A_{\mu_0} \phi_{m,n} = (m_{\mu_0})^{\frac{1}{2}} \phi_{m-\delta_{\mu\mu_0},n} \ , \quad A_{\mu_0}^* \phi_{m,n} = (m_{\mu_0}+1)^{\frac{1}{2}} \phi_{m+\delta_{\mu\mu_0},n} \qquad (4.14.2)$$

and

$$B_{\lambda_0} \phi_{m,n} = (-1)^p n_{\lambda_0} \phi_{m,n-\delta_{\lambda\lambda_0}}, \quad B_{\lambda_0}^* \phi_{m,n} = (-1)^p (1-n_{\lambda_0}) \phi_{m,n+\delta_{\lambda\lambda_0}}, \qquad (4.14.3)$$

where

$$p = \sum_{\lambda < \lambda_0} n_\lambda .$$

Here $m - \delta_{\mu\mu_0} = \{m_{\mu_1} - \delta_{\mu_1\mu_0}, \ldots, m_{\mu_k} - \delta_{\mu_k\mu_0}, \ldots\}$, with similar definitions for $m + \delta_{\mu\mu_0}$ and $n \pm \delta_{\lambda\lambda_0}$. Also, it is understood that $\phi_{m,n} = 0$ if one of the m indices is negative or if one of the n indices is different from 0 or 1.

It is clear that

$$A_{\mu_0}^* A_{\mu_0} \phi_{m,n} = m_{\mu_0} \phi_{m,n} \quad \text{and} \quad B_{\lambda_0}^* B_{\lambda_0} \phi_{m,n} = n_{\lambda_0} \phi_{m,n}$$

and that

$$(\phi_{m,n}, \phi_{m,n}) = (\phi_{m-\delta_{\mu\mu_0},n}, \phi_{m-\delta_{\mu\mu_0},n}) =$$
$$= (\phi_{m,n-\delta_{\lambda\lambda_0}}, \phi_{m,n-\delta_{\lambda\lambda_0}}) = (\phi_0, \phi_0) = 1 .$$

Thus the $\phi_{m,n}$ form an orthonormal system spanning a subspace $\mathfrak{H}_0$. Clearly $\mathfrak{H}_0$ is reduced by $\mathscr{S}$ and hence, by (v), $\mathfrak{H}_0 = \mathfrak{H}$.

That each A_μ is closed (as are of course $A_\mu^*, B_\lambda, B_\lambda^*$) was shown in the proof of Theorem 4.7.1. The operators $A_\mu, A_\mu^*, B_\lambda, B_\lambda^*$ are uniquely determined to within unitary equivalence, by (4.14.2) and (4.14.3). (See Lemma 4.4.3.) Hence by Theorem 4.5.1, $P_\mu = 2^{-\frac{1}{2}} i(A_\mu^* - A_\mu)^\sim$ and $Q_\mu = 2^{-\frac{1}{2}}(A_\mu^* + A_\mu)^\sim$ are uniquely determined by (4.14.2) to within unitary equivalence. This completes the proof of Theorem 4.14.1.

In case condition (v) is not assumed, let $\mathfrak{H}_0$ denote the space of common eigenvectors of all $A_\mu^* A_\mu$ and $B_\lambda^* B_\lambda$ belonging to the eigenvalue 0 and let $\{\phi_0^{(\alpha)}\}$ denote an orthonormal basis for $\mathfrak{H}_0$. Corresponding to (4.14.1), define $\phi_{m,n}^{(\alpha)}$ by

$$\phi_{m,n}^{(\alpha)} = \prod_\mu (m_\mu!)^{-\frac{1}{2}} (A_\mu^*)^{m_\mu} \prod_\lambda (B_\mu^*)^{n_\lambda} \phi_0^{(\alpha)} \qquad (4.14.4)$$

and, for each fixed α, let $\mathfrak{H}^{(\alpha)}$ denote the space spanned by the orthonormal system $\{\phi_{m,n}^{(\alpha)}\}$. Then $\mathfrak{H}^{(\alpha)}$ reduces $A_\mu, A_\mu^*, P_\mu, Q_\mu, B_\lambda, B_\lambda^*$, and relations corresponding to (4.14.2) and (4.14.3) hold but with ϕ replaced by $\phi^{(\alpha)}$. Consequently, one has the following result.

Theorem 4.14.2. *Let $\mathscr{S}$ be a system defined by (4.13.1) satisfying (i), (ii), (iii) and (iv). Then $\mathfrak{H} = \Sigma \oplus \mathfrak{H}^{(\alpha)}$ where each $\mathfrak{H}^{(\alpha)}$ reduces $\mathscr{S}$; and $\mathscr{S}^{(\alpha)} = \{P_\mu^{(\alpha)}, Q_\mu^{(\alpha)}, B_\lambda^{(\alpha)}, B_\lambda^{(\alpha)*}\}$, where $\mathscr{S}^{(\alpha)} = \mathscr{S}/\mathfrak{H}^{(\alpha)}$, is irreducible and uniquely determined to within unitary equivalence.*

§ 4.15 Existence of the vacuum state

In the case of infinite systems, as occur in quantum field theory, one cannot in general conclude uniqueness to within unitary equivalence. In fact, essentially distinct representations have been given by Friedrichs [3] for systems satisfying the commutation relations. Moreover, a similar situation exists in the case of the anticommutation relations; see von Neumann [4], Friedrichs [3]. However, all such representations for systems satisfying either the commutation or anticommutation relations have been reported by Gårding and Wightman [1,2]. In this connection see the book of Segal [1], in particular pp. 29–30 and the references cited there. (Incidentally, both Segal's book and that of Streater and Wightman [1] may be mentioned for their entertaining presentations of some of the mathematics and mathematical problems of field theory.)

In the case of infinite systems $\mathscr{S}$ satisfying (i), (ii), (iii), it is possible that condition (iv) is not fulfilled, in which case $\mathscr{S}$ need not be determined to within unitary equivalence. On the other hand, in the case of a finite system, condition (iv) must be satisfied, as will be shown below.

It will be convenient first to prove the following lemma.

Lemma 4.15.1. *Let $G_1, \ldots, G_p$ be a finite set of self-adjoint operators and suppose that $G_i \smile G_j$ (that is, the corresponding spectral families commute). In addition, suppose that*

$$G_i \geqq 0 \qquad (i = 1, \ldots, p). \tag{4.15.1}$$

Then $G_1 + \ldots + G_p$ is self-adjoint.

That the assertion need not hold if (4.15.1) is not assumed can be seen for $p = 2$ by choosing G_1 to be any unbounded self-adjoint operator and $G_2 = -G_1$. Then $G_1 + G_2 = G_1 + (-G_1) = 0/\mathfrak{D}_{G_1}$, so that $G_1 + G_2$ is not closed, and hence not self-adjoint. Of course $(G_1 + G_2)^\sim = 0$ is self-adjoint.

Proof. It is clear that $\mathfrak{D}_{G_1} \cap \ldots \cap \mathfrak{D}_{G_p}$ is dense. In fact, if $G_k = \int \lambda \, dE_{k\lambda}$ and if $\beta_1, \ldots, \beta_p$ are arbitrary bounded Borel sets of the real line then

$$\mathfrak{R}_{E_1(\beta_1) \ldots E_p(\beta_p)} \subset \mathfrak{D}_{G_1} \cap \ldots \cap \mathfrak{D}_{G_p} .$$

Since the G_k form a commutative system, there exist real-valued functions $g_k(\lambda)$ on $(-\infty, \infty)$ and a spectral family $\{P_\lambda\}$ such that $G_k = \int g_k(\lambda) \, dP_\lambda$; cf., e.g., Sz.-Nagy [1], p. 66. In view of (4.15.1) it can clearly be supposed that $g_k \geq 0$. Let $g(\lambda) = \sum\limits_{k=1}^{p} g_k(\lambda)$ and put $G = \int g(\lambda) \mathrm{d}P_\lambda$. Since $g^2 \leq p \Sigma g_k^2$, it is clear that $G_1 + \ldots + G_p \subset G$ and that G is self-adjoint. Since $g_k \geq 0$ then also $\Sigma g_k^2 \leq g^2$. Since $\mathfrak{D}_{G_k} = \{x : \int g_k^2 \mathrm{d} \| P_\lambda x \|^2 < \infty\}$ with a similar relation for $\mathfrak{D}_G$ it is clear that $\mathfrak{D}_G = \mathfrak{D}_{G_1} \cap \ldots \cap \mathfrak{D}_{G_p}$ and so $G = G_1 + \ldots + G_p$. This completes the proof of the lemma.

Theorem 4.15.1. *Let $\mathscr{S}$ be a finite system defined by (4.13.1) and satisfying (i), (ii) and (iii). Then necessarily (iv) holds (so that, in particular, Theorem 4.14.2 is applicable).*

Proof. The finite set of self-adjoint operators $\{A_\mu^* A_\mu, B_\lambda^* B_\lambda\}$ can be identified with the G_k's of the Lemma. Thus the operator

$$T = \sum_{\mu=1}^{r} A_\mu^* A_\mu + \sum_{\lambda=1}^{s} B_\lambda^* B_\lambda \qquad (0 \leq r, s < \infty) \qquad (4.15.2)$$

is self-adjoint. It will be shown that

$$0 \text{ is in the point spectrum of } T, \qquad (4.15.3)$$

so that $T\phi_0 = 0$ holds for some $\phi_0 \neq 0$. But this clearly implies that $A_\mu \phi_0 = 0$ and $B_\lambda \phi_0 = 0$, so that (iv) holds.

The proof of (4.15.3) will be similar to that given in § 4.4. Clearly $T \geq 0$ and hence (since it can be assumed that $T \neq 0$), there exists some $\alpha > 0$, α in $\mathrm{sp}(T)$. Let $T = \int \lambda \mathrm{d}E_\lambda$ and $\Delta_n = (\alpha - 1/n, \alpha + 1/n)$, and choose unit vectors $f_n = E(\Delta_n)f_n$. Then $(T - \alpha I)f_n = g_n \to 0$. Also,

$$T g_n = \int_{\Delta_n} \lambda \mathrm{d}E_\lambda g_n \to 0$$

and, since

$$(T g_n, g_n) = \sum_{\mu=1}^{r} \| A_\mu g_n \|^2 + \sum_{\lambda=1}^{s} \| B_\lambda g_n \|^2 ,$$

also $A_\mu g_n \to 0$ and $B_\lambda g_n \to 0$. For a fixed μ one has

$$A_\mu g_n = A_\mu (T - \alpha I)f_n = (T - (\alpha - 1)I)A_\mu f_n \to 0 \text{ as } n \to \infty . \quad (4.15.4)$$

Also if $\lambda = k$ is fixed, $B_k g_n = B_k (T - \alpha I)f_n \to 0$, that is

$$B_k (\Sigma A_\mu^* A_\mu + (I - B_k B_k^*) + \sum_{\lambda \neq k} B_\lambda^* B_\lambda - \alpha I)f_n \to 0 . \qquad (4.15.5)$$

Since $B_k B_\lambda^* B_\lambda = - B_\lambda^* B_k B_\lambda = B_\lambda^* B_\lambda B_k$ when $\lambda \neq k$ (cf. (ii) of § 4.13) and since $B_\lambda^2 = 0$ for all λ, then (4.15.5) yields

$$(T - (\alpha - I))B_k f_n \to 0 . \qquad (4.15.6)$$

But $\Sigma \| A_\mu f_n \|^2 + \Sigma \| B_\lambda f_n \|^2 = (T f_n, f_n) \to \alpha > 0$ and hence (4.15.4) and (4.15.6) imply the existence of some sequence $\{h_n\}$, $\| h_n \| = 1$, for which

$(T-(\alpha-1)I)h_n \to 0$. Thus, if $\alpha \in \mathrm{sp}(T)$, $\alpha > 0$, then $\alpha - 1 \in \mathrm{sp}(T)$. Since $T \geq 0$, it follows as in § 4.4 that $\mathrm{sp}(T) = \{0, 1, 2, \ldots\}$ and, in particular, that (4.15.3) holds. As noted earlier, this implies (iv).

§ 4.16 Self-adjointness of $\Sigma A_\mu^* A_\mu$

In the case of a denumerable infinity of commuting operators $A_\mu^* A_\mu$ the question arises as to the precise meaning of $\Sigma A_\mu^* A_\mu$. In order to formulate a possible definition, the following lemma will be proved.

Lemma 4.16.1. *Let* $G_1, G_2, \ldots$ *be a sequence of commuting self-adjoint operators (that is, the corresponding spectral families commute). In addition suppose that*

$$G_k \geq 0 \qquad (k = 1, 2, \ldots). \tag{4.16.1}$$

Define $G = \sum_{k=1}^{\infty} G_k$ *by*

$$\mathfrak{D}_G = \left\{ x \in \bigcap_{k=1}^{\infty} \mathfrak{D}_{G_k} : y = \text{w-}\lim_{n \to \infty} \sum_{k=1}^{n} G_k x \text{ exists} \right\}, \tag{4.16.2}$$

$$Gx \equiv \left(\sum_{k=1}^{\infty} G_k \right) x = y.$$

Finally, suppose that $\mathfrak{D}_G$ *is dense. Then*

$$G \text{ is self-adjoint.} \tag{4.16.3}$$

Proof. Since the G_k form a set of commuting non-negative self-adjoint operators, there exists a spectral family $\{P_\lambda\}$ and real functions $g_k(\lambda) \geq 0$ for which $G_k = \int g_k(\lambda) dP_\lambda$. If x belongs to the set $\Gamma = \{ \ldots \}$ of (4.16.2), then

$$\left(\sum_{k=1}^{n} G_k x, x \right) = \int \left(\sum_{k=1}^{n} g_k \right) d \| P_\lambda x \|^2,$$

and, in view of Lebesgue's term by term integration theorem,

$$\int \left(\sum_{k=1}^{\infty} g_k \right) d \| P_\lambda x \|^2 = \left(\sum_{k=1}^{\infty} G_k x, x \right) < \infty.$$

Since the set Γ is dense, it follows that $\int (\Sigma g_k)^{\frac{1}{2}} dP_\lambda$, hence also $G' \equiv \int (\Sigma g_k) dP_\lambda$, is self-adjoint.

Now, if $x \in \mathfrak{D}_{G'}$, so that $\int (\Sigma g_k)^2 d \| P_\lambda x \|^2 < \infty$, then, since $g_k \geq 0$, $x \in \bigcap_{k=1}^{\infty} \mathfrak{D}_{G_k}$. Also for arbitrary $y \in \mathfrak{H}$,

$$\left(\sum_{k=1}^{n} G_k x, y \right) = \int \sum_{k=1}^{n} g_k d(P_\lambda x, y) \to \int \left(\sum_{k=1}^{\infty} g_k \right) d(P_\lambda x, y)$$

as $n \to \infty$. Hence

$$x \in \Gamma \ (= \mathfrak{D}_G) \quad \text{and} \quad Gx = \int \left(\sum_{k=1}^{\infty} g_k \right) dP_\lambda x .$$

Thus

$$G' \subset G . \tag{4.16.4}$$

Next, let $x, y \in \mathfrak{D}_G$. Then

$$(Gx, y) = \lim_{n \to \infty} \left(\sum_{k=1}^{n} G_k x, y \right) = \lim_{n \to \infty} \left(x, \sum_{k=1}^{n} G_k y \right) = (x, Gy),$$

so that G is symmetric. Since G' is self-adjoint, it follows from (4.16.4) that $G' = G$ and hence (4.16.3) holds. This completes the proof of the lemma.

In case $\{G_k\}$ is a finite sequence, then $\mathfrak{D}_G$ is dense (cf. § 4.15), and Lemma 4.16.1 implies Lemma 4.15.1.

In case $\{A_\mu\}$ is a denumerable infinity of operators one can put $G_k = A_k^* A_k$ and define the (number of particles) operator $N = \sum_{k=1}^{\infty} A_k^* A_k$ by (4.16.2). Thus, if $\mathfrak{D}_N$ is dense, then N is self-adjoint. That N, for an appropriate definition, be self-adjoint corresponds to an assumption made in quantum mechanics; see Gårding and Wightman [1], p. 617, also Döring [1], p. 362.

It can be mentioned that the argument used in the proof of Theorem 4.15.1 can be used also when T is self-adjoint, even in the case of an infinite system $\mathscr{S}$. Also it is clear that if $\{B_\lambda, B_\lambda^*\}$ is a finite system then $\sum B_\lambda^* B_\lambda$ is a bounded self-adjoint operator and hence T of (4.15.2) is self-adjoint whenever $\sum A_\mu^* A_\mu$ is self-adjoint.

§ 4.17 Remarks on commutators and the equations of motion

In quantum mechanics, commutators appear in the commutation relations

$$i(PQ - QP) = I \tag{4.17.1}$$

and also in the equations of motion $dQ/dt = P$, $dP/dt = -\partial V/\partial Q$, which can be expressed as

$$i(HQ - QH) = P , \quad i(HP - PH) = -\partial V/\partial Q, \tag{4.17.2}$$

where $H = H(P, Q)$ is the Hamiltonian; cf., e.g., Ludwig [1].

Wigner [1] investigated the question as to when the relation (4.17.2) implies (4.17.1) with a Hamiltonian of the form $H = \frac{1}{2}P^2 + V(Q)$, for a suitable choice of constants. He showed that the implication fails to hold in the case of the free particle ($V = 0$) or that of the linear oscillator

$(V=Q^2)$, but that the implication is valid if $V=aQ+b$ or $V=aQ^3$ $(a\neq0)$. This last result was generalized by Putnam [3] to the case where $V=aQ^n+b(a\neq0)$ and n is any odd positive integer, with P, Q being regarded as formal elements satisfying roughly the axioms of an algebra. In this connection and concerning possible anomalies with unbounded matrices, see the remarks of Putnam [3], the reference there to Wintner [2], p. 131, also Putnam [5].

Chapter V

Wave operators and unitary equivalence of self-adjoint operators

§ 5.1 Introduction and a basic theorem

Let H_0 and H_1 be self-adjoint operators on an arbitrary Hilbert space $\mathfrak{H}$ and define the one-parameter family of unitary operators U_t by

$$U_t = U_t(H_1, H_0) = e^{itH_1} e^{-itH_0} \quad (-\infty < t < \infty). \tag{5.1.1}$$

In quantum mechanics, H_0 and H_1 correspond to the unperturbed and total Hamiltonian respectively and U_t transforms the state at time t (interaction picture) into that at time $t = 0$ (Heisenberg picture); see, e.g., Friedrichs [3], Jauch [1] and Jauch and Zinnes [1].

Let $P_j (j = 0, 1)$ denote the projection on the absolutely continuous subspace $\mathfrak{H}_a(H_j)$ of H_j (see § 2.2). If either of the strong limits W_+ or W_-, where

$$W_\pm = W_\pm(H_1, H_0) = \text{s-lim}_{t \to \pm\infty} U_t(H_1, H_0) P_0, \tag{5.1.2}$$

exists, then W_+ or W_- is called a (generalized) wave operator or half-scattering operator. In case both W_+ and W_- exist the (generalized) scattering operator is defined by

$$S = W_+^* W_- = W_+(H_1, H_0)^* W_-(H_1, H_0). \tag{5.1.3}$$

In case H_0 is absolutely continuous, that is, if $P_0 = I$, this definition is the usual one of quantum mechanics.

In this chapter there will be derived a number of results concerning the existence and properties of the wave operators.

The role of certain forms of the operators (5.1.2) ("wave matrices") in scattering theory was apparently first noted in the literature by Møller [1]. For a discussion of the quantum mechanical three particle system and for a number of references to scattering, see Faddeev [1]. A comprehensive survey of the perturbation theory of scattering can be found in the report of Krein [4].

The following was given by Kuroda [2].

Theorem 5.1.1. *Let H_j $(j=0, 1)$ be self-adjoint operators with spectral resolutions*

$$H_j = \int \lambda \, dE_{j\lambda} . \qquad (5.1.4)$$

If $W_+ = W_+ (H_1, H_0)$ exists it is a partial isometry with initial set $\mathfrak{H}_a(H_0)$ and final set $W_+ \mathfrak{H}$ contained in $\mathfrak{H}_a(H_1)$, that is

$$W_+^* W_+ = P_0 , \quad W_+ \mathfrak{H} \subset \mathfrak{H}_a(H_1) . \qquad (5.1.5)$$

Also $W_+ \mathfrak{H}$ reduces H_1 and $H_1/W_+ \mathfrak{H}$ is unitarily equivalent to H_{0a} $(=H_0/\mathfrak{H}_a(H_0))$ by

$$E_{1\lambda} W_+ = W_+ E_{0\lambda} , \quad -\infty < \lambda < \infty . \qquad (5.1.6)$$

Furthermore,

$$W_+ \mathfrak{H} = \mathfrak{H}_a(H_1) \text{ if and only if also } W_+ (H_0, H_1) \text{ exists,} \qquad (5.1.7)$$

in which case,

$$W_+ (H_1, H_0)^* = W_+ (H_0, H_1) . \qquad (5.1.8)$$

If H_2 is also self-adjoint then there holds the transitivity property that $W_+ (H_2, H_0)$ exists and is given by

$$W_+ (H_2, H_0) = W_+ (H_2, H_1) W_+ (H_1, H_0) , \qquad (5.1.9)$$

whenever the factors on the right exist. Similar results hold with W_+ replaced by W_-.

Corollary. *Suppose that both operators $W_\pm (H_1, H_0)$ exist and that $W_+ \mathfrak{H} = W_- \mathfrak{H}$. Then the scattering operator S of (5.1.3) is partially isometric with the subspace $\mathfrak{H}_a(H_0)$ as both the initial set and final set. Moreover the restriction of S to $\mathfrak{H}_a(H_0)$ is in this space a unitary operator. Hence, in case H_0 is absolutely continuous $(\mathfrak{H}_a(H_0) = \mathfrak{H})$ then S is unitary. Also S commutes with $H_0 P_0$.*

Proof. Clearly $U_t P_0$ is partially isometric with initial set $\mathfrak{H}_a(H_0)$. Since $W_+ = s\text{-}\lim_{t\to\infty} U_t P_0$ exists then $\|W_+ x\| = \lim_{t\to\infty} \|U_t P_0 x\| = \|P_0 x\|$ for all $x \in \mathfrak{H}$ and so W_+ is partially isometric with initial set $\mathfrak{H}_a(H_0)$. Since P_0 commutes with e^{itH_0}, then $e^{itH_1} U_s P_0 = U_{s+t} P_0 e^{itH_0}$. If t is fixed and $s\to\infty$, then $e^{itH_1} W_+ = W_+ e^{itH_0}$ and hence (5.1.6) holds. (Cf. Riesz and Sz.-Nagy [1], pp. 380–383.) That $W_+ \mathfrak{H}$ reduces H_1 and that $H_1/W_+ \mathfrak{H}$ is unitarily equivalent to H_{0a} is clear from (5.1.6). Since H_{0a} is absolutely continuous then so is $H_1/W_+ \mathfrak{H}$ and hence $W_+ \mathfrak{H} \subset \mathfrak{H}_a(H_1)$. This proves the assertions of Theorem 5.1.1 through formula line (5.1.6).

Next, there will be proved the transitivity property (5.1.9). It follows easily from the product rule $s\text{-}\lim (A_n B_n) = (s\text{-}\lim A_n)(s\text{-}\lim B_n)$ for strong limits and the fact that P_1 commutes with e^{itH_1} that

$$W_+(H_2, H_1)W_+(H_1, H_0) = s\text{-}\lim_{t \to \infty} e^{itH_2} P_1 e^{-itH_0} P_0. \qquad (5.1.10)$$

It must now be shown that the right side is unchanged if the factor P_1 is removed. To this end, note that $W_+(H_1, H_0)\mathfrak{H} \subset \mathfrak{H}_a(H_1)$ and, since $P_0 = W_+^*(H_1, H_0)W_+(H_1, H_0)$, also $\|W_+(H_1, H_0)x\| = \|P_0 x\|$ for all $x \in \mathfrak{H}$. Similar assertions hold for the pair (H_2, H_1). Consequently

$$\begin{aligned}
\|W_+(H_2, H_1)W_+(H_1, H_0)x\| &= \|P_1 W_+(H_1, H_0)x\| \\
&= \|W_+(H_1, H_0)x\| = \|P_0 x\|,
\end{aligned}$$

and hence by (5.1.10),

$$\lim_{t \to \infty} \|P_1 e^{-itH_0} P_0 x\| = \lim_{t \to \infty} \|e^{itH_2} P_1 e^{-itH_0} P_0 x\| = \|P_0 x\|.$$

$$(5.1.11)$$

Also,

$$\begin{aligned}
\|P_1 e^{-itH_0} P_0 x\|^2 + \|(I - P_1)e^{-itH_0} P_0 x\|^2 &= \|e^{-itH_0} P_0 x\|^2 \\
&= \|P_0 x\|^2,
\end{aligned}$$

and hence, by (5.1.10) and the fact that e^{itH_2} is unitary,

$$\lim_{t \to \infty} \|e^{itH_2}(I - P_1)e^{-itH_0} P_0 x\|^2 = 0. \qquad (5.1.12)$$

The desired relation (5.1.9) now follows from (5.1.10).

Finally, it will be proved that $W_+\mathfrak{H} = \mathfrak{H}_a(H_1)$ holds if and only if $W_+(H_0, H_1)$ exists, in which case (5.1.8) holds. First, suppose that $W_+(H_0, H_1)$ exists. Since $W_+(H_1, H_1) = P_1$, it follows from (5.1.9) and (5.1.5) that

$$\begin{aligned}
\mathfrak{H}_a(H_1) &= P_1 \mathfrak{H}_a(H_1) = \\
&= W_+(H_1, H_0)W_+(H_0, H_1)\mathfrak{H}_a(H_1) \subset W_+(H_1, H_0)\mathfrak{H} \subset \mathfrak{H}_a(H_1).
\end{aligned}$$

This implies that $W_+(H_1, H_0)\mathfrak{H} = \mathfrak{H}_a(H_1)$ and so the "if" part of (5.1.7) is proved. Next, suppose that $W_+\mathfrak{H} = \mathfrak{H}_a(H_1)$. Then for any x in $\mathfrak{H}_a(H_1)$ there exists a y in $\mathfrak{H}_a(H_0)$ such that

$$x = s\text{-}\lim_{t \to \infty} U_t(H_1, H_0)y.$$

Consequently, using the product rule for strong limits,

$$y = s\text{-}\lim_{t \to \infty} U_t(H_0, H_1)x$$

exists when $x \in \mathfrak{H}_a(H_1)$ and so

$$W_+(H_0, H_1) = s\text{-}\lim_{t \to \infty} U_t(H_0, H_1)P_1$$

exists. This proves the "only if" part of (5.1.7). But

$$W_+(H_0, H_1)W_+(H_1, H_0) = W_+(H_0, H_0) = P_0\,,$$

and since $W_+(H_0, H_1)$ is partially isometric with initial set $\mathfrak{H}_a(H_1)$, which is also the final set of $W_+(H_1, H_0)$, relation (5.1.8) follows. (For the properties of a partial isometry see § 4.12.) This completes the proof of Theorem 5.1.1.

§ 5.2 Schmidt and trace classes

If $\mathfrak{H}$ is any Hilbert space, if $B = B(\mathfrak{H})$ denotes the Banach algebra of bounded operators A on $\mathfrak{H}$, and if $\{\phi_\alpha\}$ is a complete orthonormal system in $\mathfrak{H}$, define for any $A \in B$, $\|A\|_1$ and $\|A\|_2$ by

$$\|A\|_1 = \sum_\alpha ((A^*A)^{\frac{1}{2}}\phi_\alpha, \phi_\alpha)\,, \quad \|A\|_2 = \left(\sum_\alpha \|A\phi_\alpha\|^2\right)^{\frac{1}{2}}, \quad (5.2.1)$$

so that $\|A\|_1 = \|[(A^*A)^{\frac{1}{2}}]^{\frac{1}{2}}\|_2^2$. Then $\|A\|_1$ and $\|A\|_2$ are independent of the orthonormal system chosen and are called the trace and Schmidt norms of A respectively. The trace class T consists of all bounded operators A for which $\|A\|_1 < \infty$ and the Schmidt class S of all bounded operators A for which $\|A\|_2 < \infty$. If $\|A\|$ denotes the usual norm, $\|A\| = \sup \|Ax\|$, where $\|x\| = 1$, then there hold the following inequalities:

$$\|A\| \leq \|A\|_2 \leq \|A\|_1, \ \|AB\|_i \leq \|A\| \ \|B\|_i \ \text{ and } \ \|AB\|_i \leq \|A\|_i \ \|B\|$$

$$\text{for } i = 1 \text{ and } 2,$$

$$\|AB\|_1 \leq \|A\|_2 \|B\|_2, \ \|A^*\| = \|A\|, \ \|A^*\|_i = \|A\|_i$$

$$\text{for } i = 1 \text{ and } 2.$$

If C denotes the class of completely continuous operators then $T \subset S \subset C \subset B$ and T, S, C form two-sided ideals in B. Furthermore, it is clear from (5.2.1) that $A \in T$ if and only if $(A^*A)^{\frac{1}{2}} (= [(A^*A)^{\frac{1}{2}}]^{\frac{1}{2}}) \in S$. If $A \in T$ and if A is also self-adjoint then $\|A\|_1 = \sum_{k=1}^{\infty} |\lambda_k|$ where $\{\lambda_k\}$ denotes the sequence of non-zero eigenvalues of A, and an eigenvalue is counted according to its multiplicity.

The above discussion of the trace and Schmidt norms is based on Kuroda [3], p. 249; see also Schatten [1].

§ 5.3 Some lemmas

The material of this section is based on Kato [2,3], Kuroda [3], Rosenblum [2].

Lemma 5.3.1. *Suppose that H_0 and H_1 are self-adjoint and that U_t of (5.1.1) satisfies*

$$\|(U_t - U_s)x\| \leq C_x[\eta(t;x) + \eta(s;x)] \quad (5.3.1)$$

for all $x \in \Omega$, where Ω is a dense subset of $\mathfrak{H}_a(H_0)$, C_x denotes a constant *(depending on x but independent of t, s), and $\eta(t;x) \to 0$ as $t \to \infty$. Then $W_+ = W_+(H_1, H_0)$ exists. (Similar results hold for W_-.)*

Proof. Since $s\text{-}\lim_{t\to\infty} U_t x$ exists for all $x \in \Omega$ and $\|U_t\| = 1$, it follows from the hypothesis on Ω that $s\text{-}\lim_{t\to\infty} U_t x$ exists for all $x \in \mathfrak{H}_a(H_0)$. Hence W_+ exists.

Lemma 5.3.2. *Let $H = \int \lambda \, dE_\lambda$ be self-adjoint and suppose that $\|E_\lambda x\|^2$ is absolutely continuous and satisfies*

$$\text{ess sup } d\|E_\lambda x\|^2/d\lambda \equiv m_x^2 < \infty . \tag{5.3.2}$$

Then for any $A \in S$,

$$\int_{-\infty}^{\infty} \|A e^{-itH} x\|^2 \, dt \leq 2\pi m_x^2 \|A\|_2^2 . \tag{5.3.3}$$

Proof. It is clear that it can be assumed that H is absolutely continuous. Let $\{\phi_k\}$ denote an (at most countable) orthonormal basis for the (separable) space $\mathfrak{R}_A$. Then

$$\|A e^{-itH} x\|^2 = \sum_k |(A e^{-itH} x, \phi_k)|^2$$

$$= \sum_k |\int e^{-it\lambda} [d(E_\lambda x, A^* \phi_k)/d\lambda] \, d\lambda|^2 .$$

Since

$$|d(E_\lambda x, A^* \phi_k)/d\lambda| \leq [d\|E_\lambda x\|^2/d\lambda]^{\frac{1}{2}} [d\|E_\lambda(A^* \phi_k)\|^2/d\lambda]^{\frac{1}{2}}$$

$$\leq m_x [d\|E_\lambda A^* \phi_k\|^2/d\lambda]^{\frac{1}{2}} \in L^2(-\infty, \infty)$$

it follows from a standard Lebesgue term by term integration theorem and the Fourier transform Parseval relation that

$$\int_{-\infty}^{\infty} \|A e^{-itH} x\|^2 \, dt \leq 2\pi m_x^2 \sum_k \int_{-\infty}^{\infty} [d\|E_\lambda A^* \phi_k\|^2/d\lambda] d\lambda$$

$$= 2\pi m_x^2 \sum_k \|A^* \phi_k\|^2$$

$$\leq 2\pi m_x^2 \|A^*\|_2^2 = 2\pi m_x^2 \|A\|_2^2, \text{ that is, (5.3.3)} .$$

Lemma 5.3.3. *Let H and x be defined as in Lemma 5.3.2, let $A_n \in S$ $(n = 1, 2, \ldots)$, $A \in S$ and suppose that $\|A_n - A\|_2 \to 0$ as $n \to \infty$. Then for every fixed pair s, t satisfying $-\infty \leq s \leq t \leq \infty$.*

$$\lim_{n\to\infty} \int_s^t \|A_n e^{-i\alpha H} x\|^2 \, d\alpha = \int_s^t \|A e^{-i\alpha H} x\|^2 \, d\alpha . \tag{5.3.4}$$

Proof. It follows from Lemma 5.3.2 that

$$f_n(\alpha) = \|A_n e^{-i\alpha H} x\| \quad \text{and} \quad f(\alpha) = \|A e^{-i\alpha H} x\|$$

are of class $L^2\,(-\infty,\,\infty)$ and so

$$\int_s^t |f(\alpha)-f_n(\alpha)|^2\,d\alpha \leq \int_s^t \|(A_n-A)\,e^{-i\alpha H}x\|^2\,d\alpha$$

$$\leq 2\pi m_x^2\,\|A_n-A\|_2^2\to 0 \quad\text{as } n\to\infty\,,$$

Thus $f_n\to f$ in $L^2\,(s,t)$ and (5.3.4) follows from the continuity of the norm in $L^2\,(s,t)$.

Lemma 5.3.4. *Let H_0 and V be self-adjoint, $H_1=H_0+V,\ V\in T$ (so that, in particular, H_1 is self-adjoint) and suppose that $W_+=W_+(H_1,H_0)$ exists. Let $\{E_{0\lambda}\}$ be the spectral family of H_0 and define Ω by*

$$\Omega = \{x\in\mathfrak{H}_a(H_0): \operatorname{ess\ sup}\,[d\|E_{0\lambda}x\|^2/d\lambda]\equiv m_x^2<\infty\}. \qquad (5.3.5)$$

Then Ω is dense in $\mathfrak{H}_a(H_0)$ and (5.3.1) holds with

$$C_x = (8\pi m_x^2\,\|V\|_1)^{\frac14}\,, \qquad\qquad (5.3.6)$$

$$\eta(t;x) = \left(\int_t^\infty \||V|^{\frac12}\,e^{-i\alpha H_0}x\|^2\,d\alpha\right)^{\frac14},\qquad -\infty<t<\infty,\qquad (5.3.7)$$

and $\eta(t;x)\to 0$ as $t\to\infty$. (Similar results hold for W_-.)

Proof. Let $x\in\mathfrak{H}_a(H_0)$ and let $x_n=\psi_n(H_0)$ where $\psi_n(\lambda)=1$ if

$$d\|E_{0\lambda}x\|^2/d\lambda \leq n \text{ and } \psi_n(\lambda)=0 \text{ otherwise}\,.$$

Then $x_n\in\Omega$ and $x_n\to x$ as $n\to\infty$. Consequently Ω is dense in $\mathfrak{H}_a(H_0)$. Since $|V|^{\frac12}$ belongs to S, the integral $\eta(t;x)$ of (5.3.7) is convergent by (5.3.3), and so $\lim_{t\to\infty}\eta(t;x)=0$. There remains then to show that (5.3.1) holds with $C_x,\ \eta(t;x)$ and Ω as given in the lemma. To this end, let $x\in\Omega$ and note that an integration of $d(U_tx)/dt=ie^{itH_1}Ve^{-itH_0}x$ yields

$$(U_t-U_s)x = i\int_s^t e^{i\alpha H_1}V\,e^{-i\alpha H_0}x\,d\alpha\,,\qquad x\in\mathfrak{H}\,,\qquad (5.3.8)$$

the integrand being strongly continuous in α. Next, let $x\in\mathfrak{H}_a(H_0)$, so that $x=P_0x$. Since $W_+x=\operatorname{s-lim}_{t\to\infty}U_tx$ exists and $\|W_+x\|=\|U_tx\|=\|x\|$ then $\|(W_+-U_t)x\|^2=2\,\operatorname{Re}((W_+-U_t)x,W_+x)$. It then follows from (5.3.8) and (5.1.6), which implies $e^{-i\alpha H_1}W_+=W_+\,e^{-i\alpha H_0}$, that

$$\|(W_+-U_t)x\|^2 = 2\,\operatorname{Re} i\int_t^\infty (V\,e^{-i\alpha H_0}x, W_+\,e^{-i\alpha H_0}x)\,d\alpha\,. \qquad (5.3.9)$$

But $V=\operatorname{sgn} V|V|^{\frac12}|V|^{\frac12}$, where $\operatorname{sgn} V$ is a partially isometric operator which commutes with V. An application of the Schwarz inequality then yields

$$\|(W_+-U_t)x\|^2\leq 2(\eta(t;x))^2\left(\int_t^\infty \||V|^{\frac12}(\operatorname{sgn} V)^*W_+\,e^{-i\alpha H_0}x\|^2\,d\alpha\right)^{\frac12},$$

$$(5.3.10)$$

where $\eta(t;x)$ is defined by (5.3.7). Since

$$\int_t^\infty \leq \int_{-\infty}^\infty \quad \text{and} \quad \| |V|^{\frac{1}{2}} (\operatorname{sgn} V)^* W_+ \|_2 \leq \| V \|_1^{\frac{1}{2}},$$

then Lemma 5.3.2 implies that

$$\|(W_+ - U_t)x\| \leq (8\pi m_x^2 \| V \|_1)^{\frac{1}{2}} \eta(t;x).$$

A similar relation with t replaced by s and the triangle inequality now imply the desired inequality (5.3.1) with C_x and $\eta(t;x)$ given by (5.3.6) and (5.3.7).

§ 5.4 One-dimensional perturbations

Theorem 5.4.1. *On a Hilbert space $\mathfrak{H}$, let H_0 and V be self-adjoint, V be of rank 1, and let $H_1 = H_0 + V$. Then $W_\pm = W_\pm(H_1, H_0)$ exist.*

Proof. The proof below is due to Kato [4] (in Japanese) and was given in English by Kuroda [3]. It is sufficient to prove the existence of W_+, since a similar argument establishes that of W_-. Furthermore it is clear that V can be represented in the form $Vx = c(x, \phi)\phi$ where c is a real constant and $\|\phi\| = 1$. The proof will proceed from special cases to the general case.

First, it will be supposed that $\mathfrak{H}_a(H_0) = L^2(-\infty, \infty)$ and that H_{0a} on $\mathfrak{H}_a(H_0)$ is the multiplication operator $(H_0 x)(\alpha) = \alpha x(\alpha)$. Let D^∞ be the subset of $L^2(-\infty, \infty)$ defined by

$$D^\infty = \{x \in C^\infty : \lim_{|\alpha| \to \infty} \alpha^n x^{(m)}(\alpha) = 0 \quad \text{for} \quad m, n = 0, 1, \ldots\}. \quad (5.4.1)$$

Then D^∞ is dense in $L^2(-\infty, \infty)$ $(= \mathfrak{H}_a(H_0))$ and D^∞ is invariant under the Fourier-Plancherel transform (cf. § 4.2).

Suppose first that $y = P_0 \phi$ is an element of D^∞. It will be shown that

$$\int_{-\infty}^\infty \| V e^{-i\alpha H_0} x \| \, d\alpha < \infty, \quad x \in L^2(-\infty, \infty), \quad (5.4.2)$$

and hence (cf. (5.3.8) and note that $P_0 x = x$) $W_\pm$ exist. But

$$\| V e^{-itH_0} x \| = |c| \, |(e^{-itH_0} x, \phi)| = |c| \, |(e^{-itH_0} P_0 x, \phi)| = |c| \, |(e^{-itH_0} x, y)|$$

and hence the integral of (5.4.2) is given by

$$|c| \int_{-\infty}^\infty \left| \int_{-\infty}^\infty e^{-i\alpha\lambda} x(\lambda) \bar{y}(\lambda) d\lambda \right| d\alpha. \quad (5.4.3)$$

Clearly if also $x \in D^\infty$ then $x\bar{y} \in D^\infty$ and, since D^∞ is invariant under the Fourier transform, the iterated integral is finite, and $s\text{-lim } U_t x$ exists. Since D^∞ is dense in $L^2(-\infty, \infty)$ it follows that $s\text{-lim}_{t \to \infty} U_t x$ exists for all

$x \in L^2(\infty, \infty) (= \mathfrak{H}_a(H_0))$, thus W_+ exists, provided $y = P_0 \phi$ is in D^∞.

In case $y = P_0 \phi \notin D^\infty$, choose a sequence $\{y_n\}$ satisfying $y_n \in D^\infty$, $y_n \to y$ (in L^2) and $\|y_n\| = \|y\|$. (That this is possible is clear.) If $\phi_n = \phi + y_n - y$ and $V_n = c(., \phi_n)\phi_n$ it is clear that $\phi_n \to \phi$ (in L^2). Since $|V|^{\frac{1}{2}} = |c|^{\frac{1}{2}}(., \phi)\phi$ and $|V_n|^{\frac{1}{2}} = |c|^{\frac{1}{2}}(., \phi_n)\phi_n$ it follows that

$$\lim_{n \to \infty} \| |V|^{\frac{1}{2}} - |V_n|^{\frac{1}{2}} \|_2 = 0. \tag{5.4.4}$$

In order to see this note that, since $y_n \in D^\infty$ and $y \notin D^\infty$, then y, y_n are independent and hence ϕ, ϕ_n are independent. Hence there exists a linear combination, ψ_n, of ϕ and ϕ_n satisfying $\|\psi_n\| = 1$ and $(\psi_n, \phi) = 0$ and, consequently, $(|V|^{\frac{1}{2}} - |V_n|^{\frac{1}{2}})z = 0$ whenever $(z, \phi) = (z, \phi_n) = 0$. This implies that

$$\| (|V|^{\frac{1}{2}} - |V_n|^{\frac{1}{2}}) \|_2{}^2 = \| (|V|^{\frac{1}{2}} - |V_n|^{\frac{1}{2}})\phi \|^2 + \| (|V|^{\frac{1}{2}} - |V_n|^{\frac{1}{2}})\psi_n \|^2 ,$$

and relation (5.4.4) now follows on noting that $\phi_n \to \phi$ and $(\psi_n, \phi) = 0$.

Since $P_0 \phi_n = y_n \in D^\infty$ it follows from the part of the theorem already proved that $W_+(H_0 + V_n, H_0)$ exists. An application of Lemma 5.3.4 then implies that

$$\| (U_t^{(n)} - U_s^{(n)})x \| \leq C_x^{(n)} [\eta^{(n)}(t; x) + \eta^{(n)}(s; x)], \tag{5.4.5}$$

where the superscript n signifies that V must be replaced by V_n in (5.3.6) and (5.3.7) and $U_t^{(n)} = e^{it(H_0 + V_n)} e^{-itH_0}$. Since $\|V_n\|_1 = \|V\|_1 (= |c|)$ then $C_x^{(n)} = C_x$. Also relation (5.4.4) and Lemma 5.3.3 imply that $\eta^{(n)}(t; x) \to \eta(t; x)$ as $n \to \infty$. If $H_1^{(n)} = H_0 + V_n$ then $H_1^{(n)} = H_1 + (V_n - V)$ and hence (cf. (5.3.8))

$$e^{itH_1^{(n)}} e^{-itH_1} - I = i \int_0^t e^{i\alpha H_1^{(n)}} (V_n - V) e^{-i\alpha H_1} d\alpha . \tag{5.4.6}$$

But $\phi_n \to \phi$ implies $\|V - V_n\| \to 0$ and hence $\|e^{itH_1^{(n)}} - e^{itH_1}\| \to 0$ as $n \to \infty$. Consequently $\|U_t^{(n)} - U_t\| \to 0$ as $n \to \infty$ and so (5.3.1) holds with Ω, C_x and $\eta(t; x)$ given by (5.3.5), (5.3.6) and (5.3.7). The existence of $W_+(H_1, H_0)$ then follows from Lemma 5.3.1.

Next, suppose that $\mathfrak{H}_a(H_0)$ is represented by $L^2(S)$ where S is a Borel set of real numbers and that H_{0a} is the coordinate multiplication operator on this space. Put $\mathfrak{M} = L^2(-\infty, \infty) \ominus L^2(S)$ and let $\mathfrak{H}' = \mathfrak{M} \oplus \mathfrak{H}$. Extend the original operators to the larger Hilbert space $\mathfrak{H}'$ as follows: let H_0' be the direct sum of the coordinate multiplication operator and H_0, and let V' be the direct sum of 0 and V on the respective spaces $\mathfrak{M}$ and $\mathfrak{H}$. Finally, let $H_1' = H_0' + V'$. Clearly $\mathfrak{H}_a'(H_0') = L^2(-\infty, \infty)$ and H_{0a}' is the coordinate multiplication operator on this space; also V' is of rank 1 on $\mathfrak{H}'$. Hence, by what has already been proved, $W_+(H_1', H_0')$ exists. Since $\mathfrak{H}$ reduces H_0' and V' it is clear that $W_+(H_1, H_0)$ must also exist.

Finally, consider the general case and note, as before, that $Vx = c(x, \phi)\phi$ for $x \in \mathfrak{H}$. Let $\mathfrak{H}_0$ denote the least subspace of $\mathfrak{H}$ containing ϕ and reducing H_0. Clearly $\mathfrak{H}_0$ reduces V, hence also H_1 and U_t, and moreover $V = 0$ on $\mathfrak{H} \ominus \mathfrak{H}_0$. Hence the restriction of U_t to $\mathfrak{H} \ominus \mathfrak{H}_0$ is the identity. It is therefore sufficient to prove the existence of $W_+(K_1, K_0)$ where K_0 and K_1 denote the restrictions of H_0 and H_1 to $\mathfrak{H}_0$. Clearly $\mathfrak{H}_0$ is the space generated by the vectors $\{E_{0\lambda}\phi\}$, where $\{E_{0\lambda}\}$ is the spectral family of H_0, so that one is dealing with the case of simple spectra; cf. Stone [1], Chapter 7. In fact, explicitly, let $d\sigma(\lambda) = d\|E_{0\lambda}\phi\|^2$, in the Stieltjes sense, so that one has a one-to-one correspondence

$$x = f(K_0)\phi \leftrightarrow f(\lambda) \in L^2((-\infty, \infty); d\sigma)$$

and thus a spectral representation for K_0. Clearly for K_{0a} one has the one-to-one correspondence $x = f(K_{0a})\phi \leftrightarrow f(\lambda)[d\sigma_a(\lambda)/d\lambda]^{\frac{1}{2}} \in L^2(S; d\lambda) = L^2(S)$, where $\sigma_a(\lambda)$ is the absolutely continuous part (unique to within an additive constant) of $\sigma(\lambda)$. Thus K_{0a} is represented as the coordinate multiplication operator on $L^2(S)$ and the existence of $W_+(K_1, K_0)$ follows from the preceding paragraph. This completes the proof of the theorem.

Corollary. *If V is self-adjoint and of finite rank, then $W_\pm(H_1, H_0)$ exist.*

Proof. V can be expressed as $V = \sum\limits_{k=1}^{r} c_k(., \phi_k)\phi_k$, where r is the rank of V, $\{\phi_1, \ldots, \phi_r\}$ is an orthonormal system and the c_k's are real numbers different from zero. If $H^n = H_0 + \sum\limits_{k=1}^{n} c_k(., \phi_k)\phi_k$, $n = 1, \ldots, r$, then $H^n - H^{n-1} = c_n(., \phi_n)\phi_n$ is self-adjoint, of rank 1, and $H_1 = H^r$. By Theorem 5.4.1 it follows that $W_\pm(H^n, H^{n-1})$ exist for $n = 1, \ldots, r$. Hence by the transitivity property of $W_\pm$ (cf. (5.1.9)) it follows that $W_\pm(H_1, H_0)$ exists and is given by

$$W_\pm(H_1, H^{r-1})W_\pm(H^{r-1}, H^{r-2}) \ldots W_\pm(H^1, H_0).$$

§ 5.5 Perturbations by operators of trace class

The next result is due to Rosenblum [2] and Kato [3].

Theorem 5.5.1. *Let H_0, V be self-adjoint operators, let $V \in T$, and let $H_1 = H_0 + V$. Then $W_\pm(H_1, H_0)$ and $W_\pm(H_0, H_1)$ exist. Hence (by Theorem 5.1.1), the absolutely continuous parts of H_0 and H_1 are unitarily equivalent.*

Proof. In view of the symmetric nature of the hypothesis it is sufficient to show that $W_\pm = W_\pm(H_1, H_0)$ exist. As before, it is enough to consider W_+ only. In case V is of finite rank it follows from the Corollary of Theorem 5.4.1 that $W_+(H_1, H_0)$ exists and hence, by Lemma 5.3.4, the relation (5.3.1) holds where Ω, C_x, $\eta(t; x)$ are given by (5.3.5), (5.3.6) and (5.3.7). In case V is not of finite rank, it is clear that it is the uniform limit of such operators and the validity of (5.3.1) in the more general case follows readily. An application of Lemma 5.3.1 then implies that $W_+(H_1, H_0)$ exists and the proof is complete.

§ 5.6 Invariance of wave operators

This terminology was introduced by Kato [7] and states essentially that the wave operators $W_\pm(\phi(H_1), \phi(H_0))$ exist and are independent of ϕ for a certain large class of functions ϕ.

A real-valued function $\phi = \phi(\lambda)$ on $(-\infty, \infty)$ will be said to be of class (M) if $(-\infty, \infty)$ can be expressed as the union of a finite number of disjoint open intervals $\{I_k\}$, $k = 1, 2, \ldots, r$, together with their end-points in such a way that on each I_k, $\phi(\lambda)$ is strictly monotone and has a continuous derivative $\phi'(\lambda) \neq 0$, and on any closed subinterval of I_k, $\phi'(\lambda)$ is of bounded variation. Such a system $\{I_k\}$ is of course not unique.

Kato [7] proved the following result.

Theorem 5.6.1. Let H_0, H_1 be self-adjoint operators on a Hilbert space $\mathfrak{H}$, where $H_1 = H_0 + V$ and $V \in \mathbf{T}$, and let $\{E_{0\lambda}\}$ denote the spectral family of H_0. If ϕ is of class (M) with an associated system $\{I_k\}$ then $W_\pm = W_\pm(H_1, H_0)$ and $W_\pm' = W_\pm(\phi(H_1), \phi(H_0))$ exist and satisfy

$$W_\pm' E_0(I_k) = W_\pm E_0(I_k) \text{ if } \phi \text{ is increasing on } I_k \qquad (5.6.1)$$

and

$$W_\pm' E_0(I_k) = W_\mp E_0(I_k) \text{ if } \phi \text{ is decreasing on } I_k, \qquad (5.6.2)$$

where $E_0(I_k) = E_{0,\beta_k-0} - E_{0,\alpha_k}$ if $I_k = (\alpha_k, \beta_k)$.

Consequently, $W_\pm' = W_\pm [W_\pm' = W_\mp]$ if ϕ is increasing [decreasing] in each I_k. Similar results hold with H_0 and H_1 interchanged.

Proof. That $W_\pm$ exist follows from Theorem 5.5.1. In view of the symmetry of the hypothesis it is sufficient to prove the existence and properties of $W_\pm' = W_\pm(\phi(H_1), \phi(H_0))$. Furthermore it will be clear from the proof that it is sufficient to consider only W_+'. Consider a fixed I_k and suppose that $\phi(\lambda)$ is increasing on I_k. Since $W_+ = W_+(H_1, H_0)$ exists it follows from Lemma 5.3.4, with $t = \infty$ and $s = 0$ in (5.3.1), that

$$\|(W_+ - I)x\| \leq (8\pi m^2 \|V\|_1)^{\frac{1}{4}} \left(\int_0^\infty \||V|^{\frac{1}{2}} e^{-itH_0} x\|^2 \, dt \right)^{\frac{1}{4}} \qquad (5.6.3)$$

where $x \in \mathfrak{H}_a(H_0)$ and $d\|E_{0\lambda}x\|^2/d\lambda \leq m^2$ a.e.

Next, let $y \in \mathfrak{H}_a(H_0)$ be such that $E_0(I_k)y = y$ and $d\|E_{0\lambda}y\|^2/d\lambda \leq m^2$. Then the set of such y's is dense in $E_0(I_k)\mathfrak{H}_a(H_0)$. (Cf. the beginning of the proof of Lemma 5.3.4 for a similar argument when I_k is $(-\infty, \infty)$.) If $x = e^{-is\phi(H_0)}y$ then $\|E_{0\lambda}x\|^2 = \|E_{0\lambda}y\|^2$ and so by (5.6.3),

$$\|(W_+ - I)e^{-is\phi(H_0)}y\| \leq (8\pi m^2 \|V\|_1)^{\frac{1}{4}} [\eta(s)]^{\frac{1}{4}} \qquad (5.6.4)$$

where, if $V = \Sigma \lambda_j(., \phi_j)\phi_j$,

$$\eta(s) = \int_0^\infty \| |V|^{\frac{1}{2}} e^{-itH_0 - is\phi(H_0)} y \|^2 \, dt = \sum_{j=1}^\infty |\lambda_j| A_j(s), \qquad (5.6.5)$$

with

$$A_j(s) = \int_0^\infty \left| \int_{-\infty}^\infty e^{-it\lambda - is\phi(\lambda)} w_j(\lambda) d\lambda \right|^2 dt \qquad (5.6.6)$$

and $w = w_j(\lambda) = d(E_{0\lambda} y, \phi_j)/d\lambda \in L^2(-\infty, \infty)$. Cf. § 5.3.

Now, each $A_j(s)$ has the form

$$A_j(s) = f_w(s) = 2\pi \| QU e^{-is\phi(H)} w \|^2, \quad w = w_j, \qquad (5.6.7)$$

where H denotes the self-adjoint coordinate multiplication operator on $L^2(-\infty, \infty)$, U is the unitary Fourier transform operator and Q is the projection of $L^2(-\infty, \infty)$ onto its subspace $L^2(0, \infty)$. It will be shown that

$$A_j(s) \to 0 \text{ as } s \to \infty \text{ for each fixed } j. \qquad (5.6.8)$$

Since $\| QU e^{-is\phi(H)} \| \le 1$ it is clearly enough to show that $f_w(s) \to 0$ for all characteristic functions $w = c(\lambda)$ of finite closed subintervals $[a, b]$ of I_k. But for $t, s > 0$, one has

$$\left| \int_{-\infty}^\infty e^{-it\lambda - is\phi(\lambda)} c(\lambda) d\lambda \right| = \left| \int_a^b i(t + s\phi'(\lambda))^{-1} \frac{d}{d\lambda} (e^{-it\lambda - is\phi(\lambda)}) d\lambda \right|$$

$$= \left| i \left\{ \frac{e^{-it\lambda - is\phi(\lambda)}}{t + s\phi'(\lambda)} \bigg|_a^b + is \int_a^b \frac{e^{-it\lambda - is\phi(\lambda)}}{(t + s\phi'(\lambda))^2} d\phi' \right\} \right|.$$

If C denotes the (positive) minimum of $\phi'(\lambda)$ on $[a, b]$ and if $V_{\phi'}$ denotes the variation of ϕ' on $[a, b]$ then this last expression is majorized by

$$2(t + sC)^{-1} + sV_{\phi'}(t + sC)^{-2} \le 2(t + sC)^{-1} + V_{\phi'}/C(t + sC)$$
$$= 2(C + V_{\phi'})/C(t + sC).$$

Hence for each $w = c(\lambda)$,

$$f_w(s) \le (\text{const.}) \int_0^\infty (t + sC)^{-2} dt \to 0 \text{ as } s \to \infty,$$

and (5.6.8) is now established for $w = w_j(\lambda) = d(E_{0\lambda} y, \phi_j)/d\lambda$. (In case ϕ is decreasing on I_k, $\int_0^\infty$ is to be replaced by $\int_{-\infty}^0$.)

It is clear from the Parseval relation that

$$A_j(s) \le 2\pi \int_{-\infty}^\infty |w_j(\lambda)|^2 d\lambda \le 2\pi m^2$$

and, since $\Sigma |\lambda_j| = \| V \|_1 < \infty$, it follows from (5.6.8) and (5.6.5) that

$\eta(s) \to 0$ as $s \to \infty$. Hence, by (5.6.4),

$$(W_+ - I) e^{-is\phi(H_0)} y \to 0 \quad \text{as} \quad s \to \infty \, .$$

Since

$$\| (W_+ - I) e^{-is\phi(H_0)} \| \leq 2$$

and since the set of y's considered is dense in $E_0(I_k) P_0 \mathfrak{H}$ it follows that

$$(W_+ - I) e^{-is\phi(H_0)} P_0 E_0(I_k) \to 0 \text{ strongly as } s \to \infty \, .$$

(Note that $P_0 E_0(I_k) = E_0(I_k) P_0$.) Multiplication by $e^{is\phi(H_1)}$ followed by an application of the relations $W_+ e^{-is\phi(H_0)} = e^{-is\phi(H_1)} W_+$ (cf. (5.1.6)) and $W_+ P_0 = W_+$ then yields

$$\text{s-lim}_{s \to \infty} e^{is\phi(H_1)} e^{-is\phi(H_0)} P_0 E_0(I_k) = W_+ E_0(I_k) \qquad (5.6.9)$$

whenever ϕ is increasing on I_k. Similarly,

$$\text{s-lim}_{s \to \infty} e^{is\phi(H_1)} e^{-is\phi(H_0)} P_0 E_0(I_k) = W_- E_0(I_k) \qquad (5.6.10)$$

if ϕ is decreasing on I_k. Since $P_0 E_{0\lambda}$ is continuous in λ, $\sum_k P_0 E_0(I_k) = P_0$ and so

$$\text{s-lim}_{s \to \infty} e^{is\phi(H_1)} e^{-is\phi(H_0)} P_0 = \sum_{k=1}^{r} W_{(\pm)} E_0(I_k) \, , \qquad (5.6.11)$$

where $W_{(\pm)} = W_+$ or W_- according as ϕ is increasing or decreasing on I_k.

Finally it can be noted that

$$\mathfrak{H}_a(H_0) = \mathfrak{H}_a(\phi(H_0)) \, . \qquad (5.6.12)$$

In fact, if $\phi(H_0) = \int \lambda \, dF_\lambda$ and if S is any Borel set then $F(S) = E_0(\phi^{-1}(S))$. Let S have Lebesgue measure 0; it is clear that $\phi^{-1}(S)$ also has measure 0 (ϕ being in (M)). Hence $F(S)x = 0$ whenever $\| E_{0\lambda} x \|^2$ is absolutely continuous. Also $F(\phi(S)) = E_0(\phi^{-1}(\phi(S))) \geq E_0(S)$. But if S has measure 0 so does $\phi(S)$ and hence, if $x \in \mathfrak{H}_a(\phi(H_0))$, $\| E_0(S)x \| \leq \| F(\phi(S))x \| = 0$. This proves (5.6.12). It now follows from (5.6.11) that $W_+(\phi(H_1), \phi(H_0))$ exists and is given by the right side of the equation. This completes the proof of Theorem 5.6.1.

Remark. If $U_j = (H_j - iI)(H_j + iI)^{-1}$ is the Cayley transform of H_j, then $U_j = e^{i\phi(H_j)}$ where $\phi(\lambda) = -2 \text{ arc cot } \lambda$. Since $\phi'(\lambda) > 0$ on $-\infty < \lambda < \infty$, then ϕ is in the class (M). There follows the existence of

$$W_\pm(\phi(H_1), \phi(H_0)) \quad \text{and also} \quad \text{s-lim}_{n \to \pm\infty} U_1^n U_0^{-n} P_0 = W_\pm(H_1, H_0) \, .$$

Theorem 5.6.2. *Let $\psi(\lambda)$ be of class (M) and suppose that $\psi(\lambda)$ is univalent (that is, its inverse $\hat{\psi}$ is single-valued). Let H_0 and H_1 be self-adjoint and suppose that $\psi(H_1) - \psi(H_0) \in T$. Then $W_\pm(H_1, H_0)$ and $W_\pm(H_0, H_1)$ exist.*

Proof. The domain Δ of $\hat{\psi}$ consists of a finite number of open intervals and a finite number of points. If the domain of $\hat{\psi}$ is extended to $(-\infty, \infty)$ by, for instance, putting $\hat{\psi}(\lambda) = \lambda$ on the complement of Δ then $\hat{\psi}$ is of class (M). Hence, by Theorem 5.6.1,

$$W_{\pm}(\hat{\psi}(\psi(H_1)), \hat{\psi}(\psi(H_0))) = W_{\pm}(H_1, H_0)$$

exist. In view of the symmetric nature of the hypothesis on H_0 and H_1 it is clear that $W_{\pm}(H_0, H_1)$ must exist also.

Corollary. *If 0 is not in the point spectrum of either H_0 or H_1 and if $H_1^{-p} = H_0^{-p} + V$ where $V \in T$ holds for some odd positive integer p then $W_{\pm}(\phi(H_1), \phi(H_0))$ and $W_{\pm}(\phi(H_0), \phi(H_1))$ exist for any ϕ in the class (M).*

Proof. The assertion follows from Theorem 5.6.2 if $\psi(\lambda) = \lambda^{-p}$ for $\lambda \neq 0$ and $\psi(0) = 0$.

In case $H_0 \geq 0$, $H_1 \geq 0$ then the even positive integers can also be allowed in the statement of the Corollary. For in this case, let $\psi(\lambda) = \operatorname{sgn} \lambda |\lambda|^{-p}$ for $\lambda \neq 0$ and $\psi(0) = 0$.

§ 5.7 Generalizations

As in Kato [7], by an approximate univalent sequence $\{\psi_n\}$ of functions will be meant a sequence ψ_n in the class (M) such that ψ_n is univalent on $(-n, n)$, $n = 1, 2, \ldots$.

Theorem 5.7.1. *For $j = 0$ and 1, let H_j be self-adjoint with the spectral family $\{E_{j\lambda}\}$. Let $\{\psi_n\}$ be an approximate univalent sequence for which $\psi_n(H_1) = \psi_n(H_0) + V_n$ where $V_n \in T$ $(n = 1, 2, \ldots)$. If $\phi \in (M)$ then the wave operators $W'_{\pm} = W_{\pm}(\phi(H_1), \phi(H_0))$ and $W_{\pm}(\phi(H_0), \phi(H_1))$ exist. In particular, $W_{\pm}(H_1, H_0)$ and $W_{\pm}(H_0, H_1)$ exist. Moreover $W'_{\pm}$ are piecewise equal either to $W_{\pm}$ or $W_{\mp}$ as in Theorem 5.6.1.*

The formulation of Theorem 5.7.1 and the proof to be given are due to Kato [7]. A similar theorem with a stronger hypothesis however was first given by Birman [3, 4]. See also the remarks at the end of this section.

Proof. It is clear that the restriction of ψ_n to $(-n, n)$ has an inverse function which can be extended to a $\hat{\psi}_n \in (M)$ as indicated in the preceding section. If $\phi_n = \phi \circ \hat{\psi}_n \circ \psi_n$ then $\phi_n(\lambda) = \phi(\lambda)$ on $(-n, n)$. It is easily proved that the class (M) is invariant under composition so that $\phi_n \in (M)$. Define for $j = 0, 1$ the self-adjoint operators

$$\psi_n(H_j) = L_{nj}, \quad (\hat{\psi}_n \circ \psi_n)(H_j) = H_{nj}, \tag{5.7.1}$$
$$\phi_n(H_j) = K_{nj} = \int \lambda \, dF_{nj\lambda}, \quad \phi(H_j) = K_j = \int \lambda \, dF_{j\lambda}.$$

But $K_{nj} = (\phi \circ \hat{\psi}_n) L_{nj}$, where $\phi \circ \hat{\psi}_n \in (M)$, and $L_{n1} = L_{n0} + V_n$, with $V_n \in T$, and hence

$$W'_{n\pm} = W_{\pm}(K_{n1}, K_{n0}) \quad \text{and} \quad W_{\pm}(K_{n0}, K_{n1}) \text{ exist}, \tag{5.7.2}$$

by Theorem 5.6.1.

For $\psi \in (M)$, $\psi(\pm \infty) = \lim_{\lambda \to \pm \infty} \psi(\lambda)$ exist, possibly as infinite limits. Hence $\phi_n(\pm \infty)$ and $(\hat{\psi}_n \circ \psi_n)(\pm \infty)$ exist. By choosing a suitable subsequence and retaining the same notation it can therefore be supposed that

$$\alpha_{\pm} = \lim_{n \to \infty} \phi_n(\pm \infty) \quad \text{and} \quad \beta_{\pm} = \lim_{n \to \infty} (\hat{\psi}_n \circ \psi_n)(\pm \infty) \quad \text{exist,}$$

possibly as infinite limits.

Next, let J be any open interval not containing $\alpha_{\pm}$ and $\phi(\pm \infty)$. The sets $S = \phi^{-1}(J)$ and $S_n = \phi_n^{-1}(J)$ are finite unions of open intervals and points. It follows from the definition (5.7.1) that

$$F_j(J) = E_j(S), \quad F_{nj}(J) = E_j(S_n) \quad \text{for} \quad j = 0, 1. \tag{5.7.3}$$

Since $\phi(\pm \infty)$ lie outside J, S is bounded; also, since $\alpha_{\pm}$ lie outside J, S_n is bounded for n sufficiently large. Choose n so large that S_n is bounded and $S \subset (-n, n)$. Since $\phi_n(\lambda) = \phi(\lambda)$ on $(-n, n)$, then $S = (-n, n) \cap S_n$. If $m > n$ is chosen so that $S_n \subset (-m, m)$, then also $S = (-m, m) \cap S_m$ and hence

$$S_m \cap S_n = S_m \cap (-m, m) \cap S_n = S \cap S_n = S.$$

Hence by (5.7.3),

$$F_{nj}(J) F_{mj}(J) = E_j(S_n) E_j(S_m) = E_j(S_n \cap S_m) = E_j(S) = F_j(J). \tag{5.7.4}$$

Next, let $x \in \mathfrak{H}_a(H_0) = P_0 \mathfrak{H}$. Then

$$e^{itK_{n1}}(I - F_{n1}(J)) e^{-itK_{n0}} P_0 F_0(J) = (I - F_{n1}(J)) e^{itK_{n1}} e^{-itK_{n0}} P_0 F_0(J)$$
$$\to (I - F_{n1}(J)) W'_{n+} F_0(J) \quad \text{strongly as } t \to \infty, \tag{5.7.5}$$

by (5.7.2). But $(I - F_{n1}(J)) W'_{n+} = W'_{n+}(I - F_{n0}(J))$ by (5.1.6) and $F_0(J) \leq F_{n0}(J)$ by (5.7.4). Hence the limit operator of (5.7.5) is 0. Since $\phi_n(\lambda) = \phi(\lambda)$ on $(-n, n)$ and $F_0(J) = E_0(S) \leq E_0((-n, n))$, it follows that

$$e^{-itK_{n0}} F_0(J) = e^{-itK_0} F_0(J). \tag{5.7.6}$$

Since P_0 commutes with $F_0(J)$ it now follows from (5.7.5), on multiplying by $e^{-itK_{n1}}$ on the left, that

$$\text{s-lim}_{t \to \infty} (I - F_{n1}(J)) e^{-itK_0} P_0 F_0(J) = 0. \tag{5.7.7}$$

A similar relation holds if n is replaced by the above $m > n$. If this last relation (with m) is multiplied on the left by $F_{n1}(J)$ and the result is added to (5.7.7) one obtains, in view of (5.7.4),

$$\text{s-lim}_{t \to \infty} (I - F_1(J)) e^{-itK_0} P_0 F_0(J) = 0. \tag{5.7.8}$$

If this relation is multiplied on the left by e^{itK_1} one obtains, using (5.7.6) and the similar relation $e^{itK_{n1}} F_1(J) = e^{itK_1} F_1(J)$,

$$\text{s-lim}_{t \to \infty} e^{itK_1} e^{-itK_0} P_0 F_0(J) \qquad (5.7.9)$$

$$= \text{s-lim}_{t \to \infty} F_1(J) e^{itK_1} e^{-itK_0} P_0 F_0(J)$$

$$= \text{s-lim}_{t \to \infty} F_1(J) e^{itK_{n1}} e^{-itK_{n0}} P_0 F_0(J)$$

$$= F_1(J) W'_{n+} F_0(J).$$

Thus, $\text{s-lim}_{t \to \infty} e^{itK_1} e^{-itK_0} x$ exists and is $F_1(J) W'_{n+} x$ whenever $x \in P_0 F_0(J) \mathfrak{H}$ and J is any open interval not containing $\alpha_{\pm}$ and $\phi(\pm \infty)$. Since these elements x are dense in $P_0 \mathfrak{H}$ it follows that $W'_+ = W_+(K_1, K_0)$ (and similarly, $W'_- = W_-(K_1, K_0)$) exists. An interchange of the roles of K_0 and K_1 completes the existence portion of the theorem. (Note that $\mathfrak{H}_a(K_0) = \mathfrak{H}_a(H_0)$; cf. the end of the proof of Theorem 5.6.1.)

It follows from (5.7.9) that $W'_+ x = F_1(J) W'_{n+} x$ for $x \in P_0 F_0(J) \mathfrak{H}$ (n sufficiently large, depending on J). But W'_+ and W'_{n+} are isometric on $P_0 \mathfrak{H}$, hence $\| W'_+ x \| = \| x \| = \| W'_{n+} x \|$ and, since $F_1(J)$ is a projection, $W'_+ x = W'_{n+} x$. A similar result holds for W'_- so that

$$(W'_{\pm} - W'_{n\pm}) F_0(J) = 0. \qquad (5.7.10)$$

In order to prove the remainder of the theorem, let I_k be an interval associated with $\phi \in (M)$. It can be supposed that $\phi' > 0$ on I_k. It will be shown that

$$(W'_{\pm} - W_{\pm}) E_0(I_k) = 0. \qquad (5.7.11)$$

Clearly it is enough to show that

$$(W'_{\pm} - W_{\pm}) E_0(I) = 0 \qquad (5.7.12)$$

for any finite subinterval I of I_k. In addition it can be assumed that $\beta_{\pm}$ lie outside I and that $\alpha_{\pm}, \phi(\pm \infty)$ lie outside the interval $\phi(I)$.

If $J = \phi(I)$, then $S = \phi^{-1}(J) \supset I$ and $E_j(I) \leq E_j(S) = F_j(J)$. Hence by (5.7.10),

$$(W'_{\pm} - W'_{n\pm}) E_0(I) = 0 \quad (n \text{ large}). \qquad (5.7.13)$$

Since $\beta_{\pm}$ and $\pm \infty$ lie outside I, similar results hold if $\phi(\lambda)$ is replaced by λ. In this case $W'_{\pm}, W'_{n\pm}$ must be replaced by $W_{\pm} = W_{\pm}(H_1, H_0)$ and $W_{n\pm} = W_{n\pm}(H_{n1}, H_{n0})$, so that

$$(W_{\pm} - W_{n\pm}) E_0(I) = 0 \quad (n \text{ large}). \qquad (5.7.14)$$

Let n be so large that $I \subset (-n, n)$. Clearly, the interval I can be expressed as a finite union of open subintervals Δ_p and points such that on each Δ_p, ψ_n is monotone. Since ψ_n is univalent on $(-n, n)$ then $\hat{\psi}_n$ is monotone on $\Delta'_p = \psi_n(\Delta_p)$. Since $\phi' > 0$ on $\hat{\psi}_n(\Delta'_p) = \Delta_p$, then $\phi \circ \hat{\psi}_n$ is monotone on Δ'_p (increasing or decreasing with $\hat{\psi}_n$). In view of the relations $K_{nj} = (\phi \circ \hat{\psi}_n)(L_{nj})$, $H_{nj} = \hat{\psi}_n(L_{nj})$ and $L_{n1} = L_{n0} + V_n$ with $V_n \in T$, Theorem 5.6.1 implies that

$$W_\pm(K_{n1}, K_{n0}) G_0(\Delta'_p) = W_\pm(L_{n1}, L_{n0}) G_0(\Delta'_p)$$

and

$$W_\pm(H_{n1}, H_{n0}) G_0(\Delta'_p) = W_\pm(L_{n1}, L_{n0}) G_0(\Delta'_p),$$

where $\{G_{0\lambda}\}$ denotes the spectral family of $L_{n0} = \psi_n(H_0)$. Since $E_0(\Delta_p) \leqq E_0(\psi_n^{-1}(\Delta'_p)) = G_0(\Delta'_p)$ these last relations imply $(W'_{n\pm} - W_{n\pm}) E(\Delta_p) = 0$. On summing over p one obtains

$$(W'_{n\pm} - W_{n\pm}) E_0(I) = 0 . \tag{5.7.15}$$

Relation (5.7.12) now follows from (5.7.13), (5.7.14) and (5.7.15). This completes the proof of Theorem 5.7.1.

Corollary. *Let H_0 and H_1 be self-adjoint and suppose that*

$$(H_1 - \zeta I)^{-p} - (H_0 - \zeta I)^{-p} \in T \tag{5.7.16}$$

for some positive integer p and for some non-real complex number ζ. Then $W_\pm(\phi(H_1), \phi(H_0))$ and $W_\pm(\phi(H_0), \phi(H_1))$ exist for any $\phi \in (M)$.

Proof. It is easily shown that if the hypothesis holds for some $\zeta = \zeta_0$ then it holds for all non-real ζ. If

$$\psi_n(\lambda) = i[(n - i\lambda)^{-p} - (n + i\lambda)^{-p}]$$

it follows that $\psi_n(H_1) - \psi_n(H_0) \in T$. Further, it is easily verified that $\{\psi_n\}$ contains an approximate univalent subsequence, and the Corollary now follows from Theorem 5.7.1.

A special case of the Corollary is that if

$$(H_1 - \zeta I)^{-1} - (H_0 - \zeta I)^{-1} \in T \tag{5.7.17}$$

for some ζ in the resolvent set of both H_0 and H_1 then $W_\pm(H_1, H_0)$ and $W_\pm(H_0, H_1)$ exist. In particular, by Theorem 5.1.1, these latter operators effect the unitary equivalence of the absolutely continuous parts of H_0 and H_1. If ζ is real and if (5.7.16) holds it already follows from Theorem 5.5.1 that the absolutely continuous parts of the self-adjoint operators $(H_1 - \zeta I)^{-1}$ and $(H_0 - \zeta I)^{-1}$ are unitarily equivalent via the associated wave operators. Since the absolutely continuous part of $(H_k - \zeta I)^{-1}$ coincides with that of H_k it is clear that the absolutely continuous parts of H_0 and H_1 are unitarily equivalent. What is not clear from Theorem 5.5.1 alone is, of course, that this equivalence can be implemented by the specific operators $W_\pm(H_1, H_0)$ or $W_\pm(H_0, H_1)$, or even that these exist.

It can be noted that Birman and Krein [1] define wave operators for pairs of unitary operators U_1, U_2. The existence of the wave operators is proved in case $U_1 - U_2 \in T$, while the existence of the corresponding wave operators for self-adjoint operators, when the difference of their resolvents is of trace class, is established via the Cayley transform. The proof is similar to that used by Kato [2, 3]. It may be further noted that Birman and Krein [1] also consider the scattering operator S and obtain some of its properties. Some of the methods and results of Krein [1, 3] dealing with trace formulas for self-adjoint and unitary operators are used. See also Krein [4].

Recently, Birman [5] has generalized the results of his papers [3,4] and has introduced a "local" definition of the wave operators in terms of Borel subsets G of the real line. Briefly, the situation is the following. The operator $W_{\pm}(H_1, H_0; G)$ is defined to be the strong limit as $t \to \pm \infty$ (if it exists) of $e^{itH_1} e^{-itH_0} E_0(G) P_0$, where $\{E_{k\lambda}\}$ is the spectral family of H_k and P_k is the projection onto the absolutely continuous subspace of H_k. The (ordered) pair H_0, H_1 is said to satisfy condition (α) if

$$H_1 E_1(G) E_0(G) - E_1(G) H_0 E_0(G) \in T \tag{α}$$

and to satisfy condition (β) on G if there exists a sequence of Borel sets $G_n \subset G$ $(n=0, 1, 2, \ldots)$ for which $G = G_0 + \bigcup\limits_{n=1}^{\infty} G_n$, where G_0 has Lebesgue measure 0 and

$$E_1(G') E_0(G_n) \text{ is completely continuous } (n = 1, 2, \ldots), \tag{β}$$

G' denoting the complement of G. If H_0, H_1 on the set G satisfy (α) and (β) then it turns out that the strong wave operators $W_{\pm}(H_1, H_0; G)$ exist. Furthermore, if on the set G both pairs H_0, H_1 and H_1, H_0 satisfy (α) and (β) (note that condition (β) does not involve the operators symmetrically) then both strong wave operators $W_{\pm}(H_1, H_0; G)$ and $W_{\pm}(H_0, H_1; G)$ exist.

§ 5.8 Applications to differential operators

First, the one-dimensional case will be considered. Let $g(x) > 0$, $g'(x), f(x)$ be continuous real-valued functions on $0 \leq x < \infty$ and suppose that the differential equation

$$L(u) = \lambda u, \quad L(u) \equiv -(gu')' + fu \tag{5.8.1}$$

is of the limit-point type (Weyl [2]), so that this equation and a boundary condition

$$u(0) \cos \alpha + u'(0) \sin \alpha = 0, \quad 0 \leq \alpha < \pi, \tag{5.8.2}$$

determine a (singular) boundary value problem. Corresponding to each such eigenvalue problem (that is, to each fixed α) is a self-adjoint operator $H = H_\alpha = \int \lambda dE_\lambda$, all H_α being extensions of a certain symmetric operator. In addition to Weyl [2], see also Coddington and Levinson [1], Kodaira [1], Titchmarsh [1]. It is known that (5.7.17) holds with $\zeta = i$ and H_0, H_1 any pair of extensions mentioned above. In fact, the difference of (5.7.17) is, in this case, an integral operator with kernel of rank 1. Thus the unitary equivalence of the absolutely continuous parts of H_0 and H_1 follows from the above results. For a special case see Putnam [10].

There exist similar applications to boundary value problems determined by singular elliptic operators under variations of the boundary surface and Dirichlet type boundary conditions in Euclidean m-space. It turns out that a pair of associated self-adjoint operators satisfies, under certain conditions, the relation (5.7.16) with p sufficiently large. See Birman [1,2].

§ 5.9 A sufficient condition for the existence of $W_\pm(H_1, H_0)$

Theorem 5.9.1. *Let H_0 and H_1 be self-adjoint operators for which $\mathfrak{D}_{H_0} \cap \mathfrak{D}_{H_1}$ is dense in $\mathfrak{H}_a(H_0)$ and let $V = H_1 - H_0$. In addition, let Ω denote a subset of $\mathfrak{H}_a(H_0)$ such that*

$$\text{the linear manifold determined by } \Omega \text{ is dense in } \mathfrak{H}_a(H_0) \qquad (5.9.1)$$

and

$$e^{-itH_0} x \in \mathfrak{D}_{H_0} \cap \mathfrak{D}_{H_1} \quad (= \mathfrak{D}_V) \qquad (5.9.2)$$

for all $x \in \Omega$ and all real t. Finally, suppose that

$$\int_{t_0}^{\infty} \| V e^{-itH_0} x \| \, dt < \infty , \quad \text{whenever } x \in \Omega , \qquad (5.9.3)$$

where t_0 may depend on x. Then $W_+(H_1, H_0)$ exists. A similar result holds for $W_-(H_1, H_0)$ if

$$\int_{t_0}^{\infty} \text{ is replaced by } \int_{-\infty}^{t_0} .$$

The above formulation is given in Kuroda [2]; the result was first given by Jauch and Zinnes [1]. A proof of it occurs between the lines above (cf. § 5.3) and will therefore be omitted. The theorem will be used below to obtain sufficient conditions for the existence of $W_\pm(H_1, H_0)$ when H_0 is the Hamiltonian given by the negative Laplacian operator and $V = H_1 - H_0$ is a suitable perturbation.

In connection with the existence of the wave operators see also Freeman [1], Chapter IV.

§ 5.10 Hamiltonian operators

Let the Hilbert space $\mathfrak{H}$ be the space $L^2(E_m)$ and put $H_0 = -\Delta$, where

$$\Delta \equiv \sum_{i=1}^{m} \partial^2/\partial x_i^2 \,.$$

In particular, when $m=1$, $H_0 = -d^2/dx^2$ $(=p^2, p=-id/dx)$. It is known that if the domain of H_0 is suitably specified then H_0 is a self-adjoint operator on $\mathfrak{H}=L^2(E_m)$. Henceforth this property will be assumed. Furthermore, if $V(x)$ is a real-valued, measurable function on E_m for which $V^2(x)$ is integrable on all bounded measurable subsets of E_m, the differential operator $H_0 + V(x)$, defined on an appropriate dense set (e.g., that consisting of those C^∞ functions each with a compact support), is symmetric and real, so that in particular, it has at least one self-adjoint extension. See, e.g., Sz.-Nagy [1], p. 41. Henceforth, H_1 will denote any fixed self-adjoint extension of $H_0 + V$.

It will be convenient to collect a few facts concerning m-dimensional Fourier transforms. In the terminology of quantum mechanics, one deals with the elements $f(x)$ (wave functions) of the configuration space $L^2(E_m)$, where $x=(x_1, \ldots, x_m)$, and their representations $\hat{f}(p)$ in the Fourier-Plancherel transform (momentum) space, where

$$\hat{f}(p) = (2\pi)^{-m/2} \int f(x) e^{-ipx} dx \,, \tag{5.10.1}$$

with $p=(p_1, \ldots, p_m)$, $px = p_1 x_1 + \ldots + p_m x_m$ and $dx = dx_1 \ldots dx_m$. It is well-known (cf., e.g., Stone [1], pp. 104, 441) that the configuration space of elements $f = f(x)$ and momentum space of elements $\hat{f} = \hat{f}(p)$ can be regarded as equivalent representations of the same Hilbert space. Corresponding to the operator $-i\partial/\partial x_j$ in the configuration space is the operation of multiplication by p_j in the momentum space.

If H_0 is defined as above then

$$\mathfrak{D}_{H_0} = \{f(x) \in L^2(E_m) : p^2 \hat{f}(p) \in L^2(E_m)\} \tag{5.10.2}$$

$(p^2 = \sum_{i=1}^{m} p_i^2)$; also

$$(H_0 f)\hat{}(p) = p^2 \hat{f}(p) \,. \tag{5.10.3}$$

In particular, H_0 is absolutely continuous (that is $\mathfrak{H}_a(H_0) = \mathfrak{H}$) and $\operatorname{sp}(H_0) = [0, \infty)$.

Clearly $\mathfrak{D}_V = \{f \in L^2(E_m) : V(x) f(x) \in L^2(E_m)\}$ and $(Vf)(x) = V(x) f(x)$. As noted above the operator $-\Delta + V$ has at least one self-adjoint extension, but may have many.

Concerning the operator $-\Delta + V$ in the one-dimensional case see, e.g., Coddington and Levinson [1], Kodaira [1], Titchmarsh [1],

Weyl [2] and, for the case of several variables, Kuroda [2] and the references there to Ikebe [1], Kato [1], Povzner [1], Stummel [1], Titchmarsh [2], Wienholtz [1].

§ 5.11 Existence of $W_{\pm}$ for the Hamiltonian case

Theorem 5.11.1. *Let $H_0 = -\Delta$ and V be defined on $\mathfrak{H} = L^2(E_m)$ as in § 5.10, so that V is square integrable on all bounded measurable sets, and let H_1 denote an arbitrary self-adjoint extension of $H_0 + V$. Suppose that for some $\varepsilon > 0$,*

$$V(x)\,(1+r)^{1+\varepsilon-m/2} \in L^2(E_m), \quad r = |x| . \tag{5.11.1}$$

Then the wave operators $W_{\pm}(H_1, H_0)$ exist.

The above result is due to Kuroda [2]. The proof to be given is also that of Kuroda, and is based on an argument used by Jauch and Zinnes [1]. It can be noted that the hypotheses on V in Theorem 5.11.1 are fulfilled whenever $V(x)$ is locally quadratically integrable and satisfies $V(x) = 0(r^{-1-\alpha})$ as $r \to \infty$ for some $\alpha > 0$. In case $m = 3$ the existence of the wave operators under these latter hypotheses was proved by Hack [1]. Also, when $m = 3$, Theorem 5.11.1 generalizes a result of Cook [1], who proved the existence of the wave operators under the assumption that $V(x) \in L^2(E_3)$, and of Jauch and Zinnes [1], who assumed that $V(x) = |x|^{-\beta}$, $1 < \beta < \frac{3}{2}$. In addition, it can be mentioned that (5.11.1) assures that $H_0 + V$ has a unique self-adjoint extension at least when $m \leq 3$. (Cf. Kuroda [2], p. 445 and the references cited there to Kato [1], Stummel [1], Wienholtz [1].)

For some related results see also Brownell [1].

Proof. The proof will depend on Theorem 5.9.1. Let the subset Ω be defined by

$$\Omega = \left\{ \phi_a(x) : \hat{\phi}_a(p) = \left(\prod_{i=1}^{m} p_i \right) e^{-p^2 - ipa} \right\}, \tag{5.11.2}$$

where $a = (a_1, \ldots, a_m)$ varies in E_m and $x = (x_1, \ldots, x_m)$, $p = (p_1, \ldots, p_m)$. It will be shown first that (5.9.1) and (5.9.2) hold, where now $\mathfrak{H}_a(H_0) = \mathfrak{H}$.

Since

$$(e^{-itH_0}\phi_a)\hat{\,}(p) = e^{-itp^2}\hat{\phi}_a(p) , \tag{5.11.3}$$

(cf. (5.10.3)) it follows that

$$(e^{-itH_0}\phi_a)(x) = (2\pi)^{-m/2} \int \hat{\phi}_a(p) e^{-itp^2 + ipx} dp \tag{5.11.4}$$

$$= (2\pi)^{-m/2} \prod_{i=1}^{m} \int_{-\infty}^{\infty} p_i \exp[-(1+it)p_i^2 + ip_i(x_i - a_i)] dp_i$$

$$= \text{const.}\,(1+it)^{-3m/2} \prod_{i=1}^{m} (x_i - a_i) \exp[-(x-a)^2/4(1+it)] .$$

If $t=0$, one obtains

$$\phi_a(x) = \phi(x-a), \text{ where } \phi(x) = \text{const.} \prod_{i=1}^m x_i e^{-\frac{1}{4}x^2}. \qquad (5.11.5)$$

Clearly $\hat\phi(p) = \prod_{i=1}^m p_i e^{-p^2} \neq 0$ a.e. in $L^2(E_m)$ and hence by Wiener's theorem (Wiener [1]) the linear manifold determined by the translations $\{\phi(x-a)\}$ is dense in $L^2(E_m)$. This proves (5.9.1).

It is clear from (5.11.2), (5.11.3) and (5.10.2) that $e^{-itH_0}\phi_a \in \mathfrak{D}_{H_0}$. Moreover, it is clear from (5.11.4) that whenever $\varepsilon > 0$, the function $(1+r)^{m/2-1-\varepsilon} e^{-itH_0}\phi_a(x)$ is bounded and hence, by (5.11.1),

$$V(x) e^{-itH_0} \phi_a(x) \in L^2(E_m),$$

that is,

$$e^{-itH_0}\phi_a(x) \in \mathfrak{D}_V.$$

Thus $e^{-itH_0}\phi_a \in \mathfrak{D}_{H_0} \cap \mathfrak{D}_V = \mathfrak{D}_{H_0} \cap \mathfrak{D}_{H_1}$, H_1 being an extension of $H_0 + V$, and (5.9.2) is proved.

In order to complete the proof it is clear from Theorem 5.9.1 that it is sufficient to show that

$$\int_{-\infty}^{\infty} \| V e^{-itH_0}\phi_a \| dt < \infty \qquad (5.11.6)$$

for all ϕ_a. It follows easily from (5.11.4) that

$$|e^{-itH_0}\phi_a(x)| \leq \text{const.} |1+it|^{-3m/2}|x-a|^m \exp\left[-(x-a)^2/4(1+t^2)\right]$$
$$= \text{const.}|1+it|^{-1-\delta}|x-a|^{-m/2+1+\delta}|2X|^{3m/2-1-\delta} e^{-X^2},$$
$$(5.11.7)$$

where $X = (x-a)/2(1+t^2)^{\frac{1}{2}}$ and δ is an arbitrary real number. Choose δ so that $0 < \delta < \frac{1}{2}$ and hence $3m/2-1-\delta > 0$ for any positive integer m. Then relation (5.11.7) implies

$$|V e^{-itH_0}\phi_a(x)| \leq M|1+it|^{-1-\delta}|V(x)||x-a|^{-m/2+1+\delta}, \qquad (5.11.8)$$

$0 < \delta < \frac{1}{2}$, where M is a constant independent of x and t. If now δ is chosen so that $0 < \delta < \min(\frac{1}{2}, \varepsilon)$, then the hypothesis (5.11.1) implies that for any fixed real t the right side of the inequality (5.11.8) is in $L^2(E_m)$. Consequently,

$$\| V e^{-itH_0}\phi_a \| \leq \text{const.}|1+it|^{-1-\delta}, \qquad (5.11.9)$$

where "const." depends only on a and so (5.11.6) holds. This completes the proof of Theorem 5.11.1.

Kuroda [2] also proves, under the assumptions of Theorem 5.11.1, that if $m=1$ or, if $m=2$ and $V(x)$ depends only on the distance $r=|x|$, then $W_+ \mathfrak{H} = W_- \mathfrak{H}$. It then follows from the Corollary of Theorem 5.1.1 that the scattering operator S is unitary. In addition, Kuroda also applies Theorem 5.11.1 to the scattering problem of a system of a finite number of particles. See also Kato and Kuroda [1].

For some other conditions on the potential function assuring the existence of the wave and scattering operators see §§ 5.12, 5.13 below, also Dollard [1], Green and Lanford [1], Kuroda [5], Prosser [1].

§ 5.12 A criterion for self-adjointness of perturbed operators

Theorem 5.12.1. *Let H_0, V be self-adjoint operators on a Hilbert space $\mathfrak{H}$ and suppose that*

$$\mathfrak{D}_{H_0} \subset \mathfrak{D}_V \quad and \quad \| Vx \| \leq a \| H_0 x \| + b \| x \|, \quad x \in \mathfrak{D}_{H_0}, \quad (5.12.1)$$

where a and b are constants satisfying $0 \leq a < 1$ and $0 \leq b$. Then the operator $H_1 = H_0 + V$ is self-adjoint.

The above theorem has been obtained by Rellich [1], Kato [1]. The proof below is that of Kato. The result and some of its variants are useful in establishing the self-adjoint nature of perturbed Hamiltonian operators for certain fairly general perturbations in quantum mechanics; see Kato [1], Stankevič [1].

Proof. Since $\mathfrak{D}_{H_0} \subset \mathfrak{D}_V$ it is clear that $H_1 = H_0 + V$ is symmetric. The theorem will be proved if it is shown that there exists some $k > 0$ for which the range of each of the operators $H_0 + V \pm ikI$ is $\mathfrak{H}$; cf. Sz.-Nagy [1], pp. 37–38.

To this end, let $k > 0$ and note that since H_0 is self-adjoint, $(H_0 + ikI)^{-1}$ is a bounded operator with range $\mathfrak{D}_{H_0}$. Hence, by (5.12.1), for any $x \in \mathfrak{H}$,

$$\| V(H_0 \pm ikI)^{-1} x \| \leq a \| H_0 (H_0 \pm ikI)^{-1} x \| + b \| (H_0 \pm ikI)^{-1} x \|$$
$$\leq a \| x \| + bk^{-1} \| x \| . \quad (5.12.2)$$

Since $0 \leq a < 1$ then $a + bk^{-1} < 1$ for k sufficiently large. In this case $\| V(H_0 \pm ikI)^{-1} \| < 1$ and hence the range of $I + V(H_0 \pm ikI)^{-1}$ is $\mathfrak{H}$. Since

$$H_0 + V \pm ikI = [I + V(H_0 \pm ikI)^{-1}] (H_0 \pm ikI) \quad (5.12.3)$$

and since $\mathfrak{R}_{H_0 \pm ikI} = \mathfrak{D}_{(H_0 \pm ikI)^{-1}} = \mathfrak{H}$, it follows that the range of $H_0 + V \pm ikI$ is $\mathfrak{H}$. This completes the proof of Theorem 5.12.1.

The next result was proved by Kato [1].

Theorem 5.12.2. *Let $\mathfrak{H} = L^2(E_m)$ and let the self-adjoint operators $H_0 = -\Delta$ and V be defined on $\mathfrak{H} = L^2(E_m)$ as in § 5.10. In addition, suppose that either (i) $V(x) \in L^2(E_m)$ or (ii) V is bounded at ∞ (in addition to being locally square integrable). Then (5.12.1) holds and so $H = H_0 + V$ is self-adjoint.*

Proof. Let Γ be defined by

$$\Gamma = \{f(x) \in L^2(E_m) : f(x) = P(x)e^{-\frac{1}{2}x^2}\} \qquad (5.12.4)$$

where $P(x)$ is a polynomial. The set Γ is dense in $\mathfrak{H}$ since it contains the complete Hermite orthogonal system. It follows from the properties of the Fourier-Plancherel transform that the representation $\hat{\Gamma}$ of Γ in the momentum space coincides with Γ, that is

$$\hat{\Gamma} = \{\hat{f}(p) \in L^2(E_m) : \hat{f}(p) = P(p)e^{-\frac{1}{2}p^2}\} . \qquad (5.12.5)$$

It will be shown that the set $(1+p^2)\hat{\Gamma}$ is also dense in $\mathfrak{H}$. For if $\hat{g} \in L^2(E_m)$ and if $\int \hat{g}(1+p^2)\hat{f}dp=0$ for all $\hat{f} \in \hat{\Gamma}$ then clearly $\hat{h}=\hat{g}(1+p^2)e^{-\frac{1}{2}p^2} \in L^2(E_m)$ and $\hat{h} \perp P(p)e^{-\frac{1}{2}p^2}$ for arbitrary polynomials P. Since this last system, after a change of variables, also contains the Hermite system, it follows that $\hat{h}=0$ a.e. and so $\hat{g}=0$ a.e. Thus $(1+p^2)\hat{\Gamma}$ is dense in $\mathfrak{H}$.

Next, let $f \in \Gamma$. Then $\hat{f} \in \hat{\Gamma}$ and

$$f(x) = (2\pi)^{-m/2} \int \hat{f}(p)e^{ipx} dp . \qquad (5.12.6)$$

If $\alpha > 0$ is fixed an application of the Schwarz inequality yields

$$|f(x)|^2 \leqq (2\pi)^{-m} \int |\hat{f}(p)|^2 (1+\alpha^4 p^4) dp \int (1+\alpha^4 p^4)^{-1} dp . \quad (5.12.7)$$

Hence, using the Parseval relation,

$$|f(x)|^2 \leqq \text{const.} \, \alpha^{-3}(\|f\|^2 + \alpha^4 \|H_0 f\|^2) , \qquad (5.12.8)$$

for all $f \in \Gamma$. In case (i) one obtains

$$\int |Vf|^2 dx \leqq \text{const.} \, \alpha^{-3}(\|f\|^2 + \alpha^4 \|H_0 f\|^2) \int V^2 dx ,$$

from which it follows that $f \in \mathfrak{D}_V$ and, if α is chosen sufficiently small, that

$$\|Vf\| \leqq a\|H_0 f\| + b\|f\| \qquad (5.12.9)$$

for $f \in \Gamma$ and for constants a, b satisfying $0 \leqq a < 1$, $0 \leqq b$.

Since $(1+p^2)\hat{\Gamma}$ is dense in $\mathfrak{H}$, so is $(I+H_0)\Gamma$. Hence for any $f \in \mathfrak{D}_{H_0}$ there exists a sequence $\{f_n\}$, $f_n \in \Gamma$, for which $(I+H_0)f_n \to (I+H_0)f$. Since $H_0 \geqq 0$, this implies $f_n \to f$ and hence $H_0 f_n \to H_0 f$. If f of (5.12.9) is replaced by $f_n - f_m$ it follows that $\{Vf_n\}$ is a Cauchy sequence and, since V is closed, $f \in \mathfrak{D}_V$ and $Vf_n \to Vf$. Thus $\mathfrak{D}_{H_0} \subset \mathfrak{D}_V$ and (5.12.9) holds for all $f \in \mathfrak{D}_{H_0}$, that is, (5.12.1) holds. The self-adjointness of $H_1 = H_0 + V$ then follows from Theorem 5.12.1.

In case (ii), the argument is similar if one expresses

$$\int |Vf|^2 \, dx \quad \text{as} \quad \int_{|x| \le R} + \int_{|x| > R}$$

for a sufficiently large R.

§ 5.13 Existence and properties of wave and scattering operators

The following was proved by Kuroda [2,3].

Theorem 5.13.1. Let H_0 and V satisfy the hypothesis of Theorem 5.12.1, so that in particular $H_1 = H_0 + V$ is self-adjoint, and suppose also that $|V|^{\frac{1}{2}} (H_0 + iI)^{-1} \in S$. Then the wave operators $W_\pm(H_1, H_0)$ and $W_\pm(H_0, H_1)$ exist.

Proof. Clearly $H_1 + iI = H_0 + iI + V$ and $(H_0 + iI)^{-1} x \in \mathfrak{D}_{H_0} \subset \mathfrak{D}_V$ for every $x \in \mathfrak{H}$. Hence

$$(H_1 + iI)(H_0 + iI)^{-1} x = x + V(H_0 + iI)^{-1} x ,$$

and an application of the (bounded) operator $(H_1 + iI)^{-1}$ then yields

$$(H_0 + iI)^{-1} - (H_1 + iI)^{-1} = (H_1 + iI)^{-1} V (H_0 + iI)^{-1}$$
$$= (H_1 + iI)^{-1} |V|^{\frac{1}{2}} \operatorname{sgn} V |V|^{\frac{1}{2}} (H_0 + iI)^{-1} .$$

In virtue of the Corollary to Theorem 5.7.1, the present theorem will be established if it is shown that the last operator belongs to T. Since $|V|^{\frac{1}{2}} (H_0 + iI)^{-1}$, hence also $\operatorname{sgn} V |V|^{\frac{1}{2}} (H_0 + iI)^{-1}$, belongs to S, it is sufficient then to show that

$$Q = (H_1 + iI)^{-1} |V|^{\frac{1}{2}} \in S \text{ or } Q^* = |V|^{\frac{1}{2}} (H_1 - iI)^{-1} \in S. \quad \text{(Cf. § 5.2.)}$$

Since

$$Q^* = |V|^{\frac{1}{2}} (H_0 + iI)^{-1} (H_0 + iI)(H_1 - iI)^{-1} \text{ and } (H_0 + iI)(H_1 - iI)^{-1}$$

is bounded, then $Q^* \in S$ and the proof is complete.

The next result is due to Kuroda [2].

Theorem 5.13.2. Let $\mathfrak{H} = L^2(E_m)$ with $m \le 3$ and let the self-adjoint operators $H_0 = -\Delta$ and V be defined on $L^2(E_m)$ as in § 5.10. In addition, suppose that

$$V(x) \in L^1(E_m) \cap L^2(E_m) . \tag{5.13.1}$$

Then $H_1 = H_0 + V$ is self-adjoint and the wave operators $W_\pm(H_1, H_0)$ and $W_\pm(H_0, H_1)$ exist. In particular, since H_0 is absolutely continuous, it follows from Theorem 5.1.1 and its Corollary, that the scattering operator S is unitary.

Proof. Since $V \in L^2(E_m)$, H_1 is self-adjoint by Theorem 5.12.2. The present theorem will follow from Theorem 5.13.1 if it is shown that $|V|^{\frac{1}{2}} (H_0 + iI)^{-1} \in S$. To this end, note that for any $f \in \mathfrak{H}$, one has

$$((H_0+iI)^{-1}f)(x) = (2\pi)^{-m/2} \int (p^2+i)^{-1} e^{ipx} \hat{f}(p)\mathrm{d}p . \qquad (5.13.2)$$

Since $m \leq 3$, $(p^2+i)^{-1} \in L^2(E_m)$. Hence, if

$$h(x) = (2\pi)^{-m} \int (p^2+i)^{-1} e^{ipx} \mathrm{d}p , \qquad (5.13.3)$$

then h is the inverse Fourier transform of $(2\pi)^{-m/2}(p^2+i)^{-1} \in L^2(E_m)$, so $h \in L^2(E_m)$ and

$$((H_0+iI)^{-1}f)(x) = \int h(x-y)f(y)\mathrm{d}y . \qquad (5.13.4)$$

Clearly,

$$(|V|^{\frac{1}{2}}(H_0+iI)^{-1}f)(x) = \int |V(x)|^{\frac{1}{2}} h(x-y)f(y)\mathrm{d}y , \qquad (5.13.5)$$

so that $|V|^{\frac{1}{2}}(H_0+iI)^{-1}$ is an integral operator with kernel $K(x, y) = |V(x)|^{\frac{1}{2}} h(x-y)$. But, in view of (5.13.1),

$$\int\int |K(x, y)|^2 \mathrm{d}x\,\mathrm{d}y = \int |V(x)| \mathrm{d}x \int |h(y)|^2 \mathrm{d}y < \infty, \qquad (5.13.6)$$

and so $K(x, y)$ is a Hilbert-Schmidt kernel. Consequently, as was to be shown, $|V|^{\frac{1}{2}}(H_0+iI)^{-1} \in S$, and the proof is complete.

Ikebe [1] has obtained a more general form of the preceding using the eigenfunction expansion theory associated with the operator $-\Delta+V$. Concerning the spectrum of this operator under hypotheses similar to those of Theorem 5.13.2 see also Povzner [1]. (See the remarks of Kuroda [2], p. 453.)

It is seen that the above two theorems of Kuroda can be applied to Schrödinger operators in $L^2(E_m)$ provided $m \leq 3$. Stankevič [1] has obtained a generalization of Theorem 5.13.1 which, in particular, extends the application to spaces of arbitrary dimension.

For some related results see Birman and Entina [1], also the survey paper of Krein [4].

Kuroda [4] considers the case where H_0 is bounded from below and the perturbed operator H_1 is defined not by $H_1 = H_0 + V$ as above, but indirectly by means of a Hermitian form bounded from below. Under certain hypotheses, the Hermitian form associated with the perturbation is of trace class with reference to an appropriate metric and the existence of the wave operators is again established.

§ 5.14 Stationary approach to scattering

So far the unitary equivalence of the absolutely continuous parts of H_0 and $H_1 = H_0 + V$ has been considered only from the point of view of the operators

$$U_{\pm} = \text{s-lim}_{t \to \pm\infty} e^{itH_1} e^{-itH_0} P_0 \text{ (when these exist).}$$

This is the time-dependent case. Another point of view is the time-independent or stationary one in which the parameter t does not enter. In the special case of a perturbation V of rank 1, Kato [2] considered this method, an outline of which is as follows.

If $\mathfrak{M}$ denotes the smallest space reducing H_0 and containing $\mathfrak{R}_V$, it is clear that $\mathfrak{M}$ also reduces V, and that $\mathfrak{M}$ is also the smallest space reducing H_1 and containing $\mathfrak{R}_V$. Moreover, H_0 and H_1 coincide in $\mathfrak{M}^\perp = \mathfrak{H} \ominus \mathfrak{M}$ and hence in this space the identity serves as a unitary transformation effecting the equivalence of the absolutely continuous parts of H_0 and H_1. Consequently, it can be assumed that $\mathfrak{M} = \mathfrak{H}$.

It can be supposed that $Vx = c(x, \phi)\phi$ where c is real and $\phi \in \mathfrak{H}$, $\phi \neq 0$. Then H_0 and H_1 have simple spectra and, if $H_j = \int \lambda dE_{j\lambda}$, there exists for each $x \in \mathfrak{H}$ a pair of Baire functions $f_j(\lambda, x)$ for which

$$x = \int f_j(\lambda, x) dE_{j\lambda}\phi \quad \text{and} \quad \|x\|^2 = \int |f_j(\lambda, x)|^2 d\|E_{j\lambda}\phi\|^2.$$

It follows from $H_1 = H_0 + V$ that

$$(H_1 - \zeta I)^{-1} - (H_0 - \zeta I)^{-1} = -(H_0 - \zeta I)^{-1} V (H_1 - \zeta I)^{-1}$$

whenever $\text{Im}(\zeta) \neq 0$ and hence

$$((H_1 - \zeta I)^{-1} x, \phi) - ((H_0 - \zeta I)^{-1} x, \phi)$$
$$= -c((H_1 - \zeta I)^{-1} x, \phi)((H_0 - \zeta I)^{-1}\phi, \phi).$$

If $\rho_j(\lambda) = \|E_{j\lambda}\phi\|^2$ and if for $\rho = \rho_j$ one defines

$$J(\zeta, f, \rho) = \int (\lambda - \zeta)^{-1} f(\lambda) d\rho(\lambda)$$

and

$$J(\zeta, \rho) = J(\zeta, 1, \rho),$$

then

$$[1 + cJ(\zeta, \rho_0)] J(\zeta, f_1(\ , x), \rho_1) = J(\zeta, f_0(\ , x), \rho_0),$$

and similarly,

$$[1 - cJ(\zeta, \rho_1)] J(\zeta, f_0(\ , x), \rho_0) = J(\zeta, f_1(\ , x), \rho_1).$$

Since $f_j(\lambda, \phi) = 1$, these equations imply for $x = \phi$ the relation

$$[1 + cJ(\zeta, \rho_0)][1 - cJ(\zeta, \rho_1)] = 1.$$

It follows from the inversion formula for Poisson-Stieltjes integrals that whenever $f \in L^2(\rho)$, the limits

$$J(\lambda \pm i0, f, \rho) = \lim_{\varepsilon \to 0+} J(\lambda \pm i\varepsilon, f, \rho) \text{ exist a.e.}$$

as finite values and that, a.e.,

$$J(\lambda+i0,f,\rho) - J(\lambda-i0,f,\rho) = 2\pi i f(\lambda)\rho'(\lambda) \,.$$

Hence if $w(\zeta) = 1 + cJ(\zeta,\rho_0) = [1 - cJ(\zeta,\rho_1)]^{-1}$, then $w(\lambda+i0)$ and $w(\lambda-i0)$ are complex conjugates and $\rho_1'(\lambda) = |w(\lambda\pm i0)|^{-2}\rho_0'(\lambda)$ a.e., where $w(\lambda\pm i0)\neq 0$. In particular, $\rho_1'(\lambda)=0$ if and only if $\rho_0'(\lambda)=0$ (a.e.) and the absolutely continuous parts of ρ_0 and ρ_1 are equivalent. The unitary equivalence of H_0 and H_1 then follows; cf., e.g., Stone [1], Halmos [3].

Clearly there exists a Borel set A which is a support of the absolutely continuous parts of both ρ_0 and ρ_1 and on which both $\rho_0'(\lambda)$ and $\rho_1'(\lambda)$ exist and are positive. Then, since $w(\lambda\pm i0)$ is defined a.e. and the spectral measure $E_{1\lambda}$ is absolutely continuous on A, one can define $W_\pm$ by

$$W_\pm x = \int_A w(\lambda\pm i0)f_0(\lambda,x)\mathrm{d}E_{1\lambda}\phi \,. \qquad (5.14.1)$$

Then

$$\|W_\pm x\|^2 = \int_A |w(\lambda\pm i0)|^2 |f_0(\lambda,x)|^2 \rho_1'(\lambda)\mathrm{d}\lambda$$

$$= \int_A |f_0(\lambda,x)|^2 \rho_0'(\lambda)\mathrm{d}\lambda = \|P_0 x\|^2 \,,$$

where P_0 denotes the projection of $\mathfrak{H}$ onto the space $\mathfrak{H}_a(H_0)$, so that $W_\pm$ are partially isometric operators with initial set $\mathfrak{H}_a(H_0)$. Similarly, one can show that

$$W_\mp^* y = \int_A w(\lambda\pm i0)^{-1}f_1(\lambda,y)\mathrm{d}E_{0\lambda}\phi$$

and that $\|W_\mp^* x\| = \|P_1 x\|$, where P_1 is the projection of $\mathfrak{H}$ onto $\mathfrak{H}_a(H_1)$. Moreover, if X is any Borel subset of A, then

$$W_\pm E_0(X)x = W_\pm \int_X f_0(\lambda,x)\mathrm{d}E_{0\lambda}\phi =$$

$$= \int_X w(\lambda\pm i0)f_0(\lambda,x)\mathrm{d}E_{1\lambda}\phi = E_1(X)W_\pm x,$$

hence $W_\pm E_0(X) = E_1(X)W_\pm$, and therefore also $E_0(X)W_\mp^* = W_\mp^* E_1(X)$. Thus $W_\pm$ are explicit unitary operators effecting the equivalence of the absolutely continuous parts of H_0 and H_1.

It can be shown further (cf. Kato [2]) that the operators $W_\pm$ of (5.14.1) coincide with the wave operators $W_\pm$ defined earlier.

Kuroda [6] extends the time-independent method to perturbations V of finite rank.

De Branges [1], using operator-valued measures and an associated integration theory, gives a time-independent proof of the unitary equivalence of the absolutely continuous parts of two self-adjoint operators H_0 and H_1 on a Hilbert space when the difference $(H_1 - \zeta I)^{-1} - (H_0 - \zeta I)^{-1}$ for some (hence every) non-real ζ belongs to trace class. He proves the existence of non-negative operator valued measures μ_0 and μ_1 defined for bounded Borel sets of the real line and satisfying

$$\int (1 + t^2)^{-1} \, d\tau(\mu_k(t)) < \infty \, ,$$

where τ denotes the trace norm, along with explicitly constructed linear isometric transformations V_k of $\mathfrak{H}$ onto Hilbert spaces $L^2(\mu_k)$ of vector-valued functions, in such a way that each V_k effects the unitary equivalence of H_k on $\mathfrak{H}$ with multiplication by x in $L^2(\mu_k)$. For every f in $\mathfrak{H}$ the element $V_0 f$ of $L^2(\mu_0)$ is a generalized Hilbert transform of $V_1 f$ in $L^2(\mu_1)$.

Kuroda [7] establishes a connection between the operators of the time-independent and the time-dependent scattering theory. His method depends upon a lemma involving the limits of certain operators in the form of integrals, which was given by de Branges [1], and also on the methods used by Kato (see above) for perturbations of rank 1.

Another stationary approach to scattering, by a method involving limits of resolvent integrals, is given by Birman and Entina [1]. Briefly, their method is the following. Let $H_1 = H_0 + V$ where all operators are self-adjoint and V is of trace class and let $\{E_{k\lambda}\}$ denote the spectral family of H_k. If

$$R_k(\zeta) = (H_k - \zeta I)^{-1} \quad \text{and} \quad Q_0(\zeta) = I + V R_0(\zeta), \, Q_1(\zeta) = I - V R_1(\zeta)$$

then $Q_0(\zeta) = Q_1^{-1}(\zeta)$ and

$$R_0(\zeta) - R_0(\bar{\zeta}) = Q_0^*(\zeta)(R_1(\zeta) - R_1(\bar{\zeta})) Q_0(\zeta) \, . \tag{5.14.2}$$

It turns out that the strong limits

$$Q_k(\lambda \pm i0) f = \operatorname*{s-lim}_{\varepsilon \to 0+} Q_k(\lambda \pm i\varepsilon) f$$

exist for almost all λ and that the derivatives

$$d(E_{1\lambda} Q_0(\mu \pm i0) f, g)/d\lambda \quad \text{and} \quad d(E_{1\lambda} Q_0(\mu \pm i0) f, Q_0(\mu \pm i0) g)/d\lambda$$

exist for almost all λ and μ, where f and g are arbitrary. Further, as a consequence of (5.14.2), the latter satisfies

$$[d(E_{1\lambda} Q_0(\mu \pm i0) f, Q_0(\mu \pm i0) g) d\lambda]_{\mu = \lambda} = d(E_{0\lambda} f, g)/d\lambda \, .$$

It then follows that if P_k denotes the orthogonal projection of $\mathfrak{H}$ onto $\mathfrak{H}_a(H_k)$ then

$$\int_{-\infty}^{\infty} |[d(E_{1\lambda} Q_0(\mu \pm i0) f, g) d\lambda]_{\mu = \lambda}| d\lambda \leq \| P_0 f \| \, \| P_1 g \| \, .$$

If the bounded operators $W_\pm$ are then defined by

$$(W_\pm f, g) = \int_{-\infty}^{\infty} [d(E_{1\lambda} Q_0(\mu \pm i0)f, g)/d\lambda]_{\mu = \lambda} d\lambda , \qquad (5.14.3)$$

it turns out that $W_\pm$ coincide with the wave operators considered earlier and are also given by

$$(W_\pm f, g) = \int_{-\infty}^{\infty} [d(E_{0\lambda} Q_1(\mu \pm i0)f, g)/d\lambda]_{\mu = \lambda} d\lambda . \qquad (5.14.4)$$

An integral similar to (5.14.3) also occurs in Birman [2], p. 410. Birman and Entina [1] also obtain representations similar to (5.14.3) and (5.14.4) for the scattering operator $S = W_+^* W_-$.

In a recent paper, Kato [8] develops a stationary (time-independent) perturbation theory involving an explicit construction of "wave operators" implementing the similarity of not necessarily self-adjoint operators.

§ 5.15 Non-negative perturbations

As above, let H_0, H_1 and V denote self-adjoint operators satisfying $H_1 = H_0 + V$. It has been shown above that for certain types of perturbations V, the absolutely continuous parts of H_0 and H_1 are unitarily equivalent. Among the earliest theorems of this type are those of Friedrichs [1,2], who considered in particular the problem where $\mathfrak{H} = L^2(-1, 1)$, H_0 was the bounded absolutely continuous operator defined by $(H_0 f)(x) = x f(x)$ and V was an integral operator with a kernel satisfying certain Lipschitz conditions. (See also the introductory remarks in Rosenblum [2].) Friedrichs showed that if $H_1 = H_1^\varepsilon$ is defined by

$$H_1^\varepsilon = H_0 + \varepsilon V, \quad \varepsilon \text{ real}, \qquad (5.15.1)$$

then the strong limits

$$U_\pm^\varepsilon = s\text{-}\lim_{t \to \pm \infty} e^{itH_1^\varepsilon} e^{-itH_0} \qquad (5.15.2)$$

exist for ε sufficiently small and effect the unitary equivalence of H_1^ε and H_0, thus

$$H_1^\varepsilon = U_+^\varepsilon H_0 U_+^{\varepsilon *}, \quad H_1^\varepsilon = U_-^\varepsilon H_0 U_-^{\varepsilon *} . \qquad (5.15.3)$$

Furthermore, each of the operators $U^\varepsilon = U_+^\varepsilon$, U_-^ε is analytic in ε and of the form

$$U^\varepsilon = I + \varepsilon U_1 + \varepsilon^2 U_2 + \ldots, \quad U^\varepsilon \text{ unitary} . \qquad (5.15.4)$$

In particular, $\| U^\varepsilon - I \| \to 0$ as $\varepsilon \to 0$ and consequently

$$\text{sp}(U^\varepsilon) \to 1 \text{ as } \varepsilon \to 0 . \qquad (5.15.5)$$

It turns out that for certain types of analytic perturbations (5.15.1), where H_0, H_1 and V are all self-adjoint, in order that a perturbation technique such as that described above (that is, one which yields a unitary family U^ε satisfying (5.15.3) and (5.15.4)) be valid, it is necessary that H_0 either be absolutely continuous or at least have an absolutely continuous part. In fact, there will be proved the following.

Theorem 5.15.1. *Let H_0 be a self-adjoint operator bounded from below, let V satisfy*

$$V \geqq 0, \quad V \neq 0, \tag{5.15.6}$$

and let H_1^ε of (5.15.1) also be self-adjoint. Then in order for (5.15.3) and (5.15.4) to be valid, for sufficiently small ε, it is necessary that H_0 have an absolutely continuous part. If, in addition, 0 is not in the point spectrum of V then it is even necessary that H_0 be absolutely continuous.

Proof. If (5.15.4) holds then meas $\mathrm{sp}(U^\varepsilon) < 2\pi$ for ε sufficiently small. It follows from Theorem 2.12.2 that if $\varepsilon > 0$ and if $\mu < \min H_0$ then $\mathfrak{H}_a(H_0)$ contains the range of the self-adjoint operator

$$(H_0 - \mu I)^{-1} V (H_1^\varepsilon - \mu I)^{-1}.$$

Since $V \neq 0$, H_0 must have an absolutely continuous part. If 0 is not in the point spectrum of V then clearly the closure of the above range is $\mathfrak{H}$ and so H_0 is absolutely continuous.

§ 5.16 Hamiltonians and non-negative perturbations

As in § 5.10, H_0 will be defined on $\mathfrak{H} = L^2(E_m)$ by $H_0 = -\Delta$ as a self-adjoint operator. In the case of one-dimension there will be proved the following result (Putnam [22]).

Theorem 5.16.1. *Let*

$$H_0 = -\mathrm{d}^2/\mathrm{d}x^2 \quad on \quad \mathfrak{H} = L^2(E_1) \tag{5.16.1}$$

and let $V(x)$ be a real-valued, measurable function satisfying

$$0 \leqq V(x) \leqq \text{const.}, \text{ a.e.} \tag{5.16.2}$$

Suppose that the self-adjoint operator $H_1 = H_0 + V$ is unitarily equivalent to H_0 and that U is any unitary operator effecting this equivalence, that is,

$$H_1 = U H_0 U^*, \quad U \text{ unitary}. \tag{5.16.3}$$

Then

$$\mathfrak{H}_a(U) \supset \mathfrak{R}_V. \tag{5.16.4}$$

If, in addition,

$$\liminf_{b-a \to \infty} (b-a)^{-3} \int_a^b V^{-1}(x)\mathrm{d}x = 0 \tag{5.16.5}$$

then

$$\mathrm{sp}(U) = \{\lambda : |\lambda| = 1\}. \tag{5.16.6}$$

Proof. Relation (5.16.4) follows from Theorem 2.8.1 with $J = H_0$, $D = V$ and $B = H_1$. Theorem 2.7.1 will be used to prove (5.16.6) under the additional hypothesis (5.16.5). It will be shown that

$$\inf_g \left\{ \int_{-\infty}^{\infty} V^{-1} g^2 \, dx \int_{-\infty}^{\infty} g'^2 \, dx \Big/ \left(\int_{-\infty}^{\infty} g^2 \, dx \right)^2 \right\} = 0, \quad (5.16.7)$$

where $g \in C^2$ and g, g'' (hence g') and $V^{-\frac{1}{2}} g$ are real and belong to $L^2(-\infty, \infty)$. Since

$$(H_0 g, g) = - \int g g'' \, dx = \int g'^2 \, dx,$$

the assertion (5.16.6) will then follow from (2.7.1).

Let $h = h(x)$ be defined on $-\infty < x < \infty$ by $h(x) = (b-a)^{-\frac{3}{2}} (x-a)$ if $a \le x \le \frac{1}{2}(a+b)$, $h(x) = h(a+b-x)$ if $\frac{1}{2}(a+b) \le x \le b$, and $h(x) = 0$ outside $[a, b]$. Then $h(x)$ is absolutely continuous,

$$\int_{-\infty}^{\infty} h^2 \, dx = 1/12, \quad \int_{-\infty}^{\infty} h'^2 \, dx = (b-a)^{-2}$$

and

$$\int_{-\infty}^{\infty} V^{-1} h^2 \, dx \le \int_a^b V^{-1} \, dx / 4(b-a).$$

Hence the expression $\{\ldots\}$ of (5.16.7) with g replaced by h satisfies

$$|\{\ldots\}| \le \text{const.} (b-a)^{-3} \int_a^b V^{-1}(x) \, dx. \quad (5.16.8)$$

Moreover it is clear that h can be smoothed out so as to obtain a function g of the type mentioned above and so that (5.16.8) remains valid. Thus (5.16.7) is established and the proof is complete.

Theorem 5.16.2. *Let H_0 be defined by (5.16.1) and suppose that $V(x)$ satisfies (5.16.2) and that $V \in L(-\infty, \infty)$. Then H_0 and H_1 are absolutely continuous and the wave operators $U_\pm = $ s-lim $e^{itH_1} e^{-itH_0}$ exist and are* $\quad t \to \pm \infty$ *unitary operators satisfying (5.16.3) and (5.16.4) with $U = U_\pm$. If also (5.16.5) is assumed then (5.16.6) holds.*

Proof. Since V is bounded, it is clear that (5.13.1) holds with $m = 1$ and so the existence of $U_\pm$ follows from Theorem 5.13.2. That H_0 is absolutely continuous was pointed out above; cf. § 5.10, also Coddington and Levinson [1], Kodaira [1], Titchmarsh [1], p. 59, Weyl [2]. Relation (5.16.2) implies that $\text{sp}(H_1) \subset [0, \infty)$. Moreover, asymptotic formulas for the solutions of $y'' + (\lambda - V(x)) y = 0$ (cf. Wintner [5], p. 421) assure that H_1 is absolutely continuous and that its spectrum is $[0, \infty)$. Hence $U_\pm$ are unitary and satisfy (5.16.3). The remaining assertions with $U = U_\pm$ then follow from Theorem 5.16.1.

It is seen that the condition $V \in L(-\infty, \infty)$ implies that V is "small" while the condition (5.16.5) implies that V is "not too small." It can be noted that in the extreme case in which $V(x) \equiv 0$ then (5.16.2) holds and $V \in L(-\infty, \infty)$, but (5.16.5) does not hold. Moreover, in this case, (5.16.6) need not hold; in fact, each wave operator is now the identity operator. It is easily verified that all of the hypotheses of Theorem 5.16.2 (including (5.16.5)) are fulfilled if, say, $V(x) = |x|^{-c}$ with $1 < c < 2$ for $|x|$ large and satisfies (5.16.2) everywhere.

If ε is real and if

$$H_1^\varepsilon = H_0 + \varepsilon V, \tag{5.16.9}$$

it is clear from the discussion of § 5.15 that if (5.16.5) is included in the hypotheses of Theorem 5.16.1 or of Theorem 5.16.2, then one must rule out the possibility of an analytic family of unitary operators U^ε of the form (5.15.4) and satisfying $H_1^\varepsilon = U^\varepsilon H_0 U^{\varepsilon*}$.

If the assumption (5.16.2) occurring in Theorems 5.16.1 and 5.16.2 is strengthened to $0 < V(x) \leq$ const. almost everywhere, then $[\mathfrak{R}_V] = \mathfrak{H}$ and it follows from (5.16.4) that U is absolutely continuous.

Theorem 5.16.3. *Let $H_0 = -\Delta$ on $L^2(E_m)$ with $m \geq 1$ and suppose further that V is a radial potential, that is, $V = V(r)$ where $r = (x_1^2 + \ldots + x_m^2)^{\frac{1}{2}}$. Suppose that*

$$0 \leq V(r) \leq \text{const.} < \infty \text{ a.e.}, \tag{5.16.10}$$

$$\liminf_{R \to \infty} R^{-3} \int_{R_0}^R V^{-1}(r) \, dr = 0, \tag{5.16.11}$$

and that (5.16.3) holds. Then (5.16.6) is true.

Proof. The proof is similar to the corresponding part of that of Theorem 5.16.1. The assertion follows from Theorem 2.7.1 if, rather than (5.16.7), one shows that

$$\inf_g \left\{ \int_0^\infty V^{-1} g^2 r^{m-1} \, dr \int_0^\infty g'^2 r^{m-1} \, dr \Big/ \int_0^\infty g^2 r^{m-1} \, dr \right\} = 0, \tag{5.16.12}$$

where $g(r) \in C^2$ and vanishes for sufficiently large r. However relation (5.16.12) is readily established by a construction similar to that used in connection with (5.16.7) and the proof is complete.

§ 5.17 Remarks on unitary equivalence

The investigations of this chapter, involving wave operators and unitary equivalence of the absolutely continuous parts of self-adjoint operators, concerns of course only one of many aspects of the theory of perturbations of self-adjoint operators.

In general, for the possibility of the unitary equivalence of two self-adjoint operators, one being regarded as a perturbed form of the other, it is clear that some restrictions must be imposed. Simple examples, even in the case of finite-dimensional Hilbert spaces, show that the spectrum and the point spectrum are hardly ever left invariant , much less unitarily invariant, under any kind of perturbation. On the other hand, Weyl [1] showed that the essential spectrum—that is, the set of limit (non-isolated) points of the spectrum together with points in the point spectrum of infinite multiplicity—is at least invariant under self-adjoint completely continuous perturbations. This fact is of considerable importance in the theory of singular boundary value problems, Weyl [2], where a change of boundary condition corresponds to a one-dimensional perturbation of the resolvent operator. See also Hartman [1], concerning the general problem of the connection between the spectra of self-adjoint extensions of a fixed symmetric operator; Wolf [1], concerning differential operators differing by a finite number of one-dimensional boundary conditions; and Wolf [2, 3, 4], Kreith and Wolf [1], on operators arising in partial differential boundary value problems.

However, although the essential spectrum thus remains stable under certain types of small perturbations (completely continuous ones) its composition may be completely altered so as to deny any possibility of unitary equivalence. It is known for instance (Weyl [1], von Neumann [5]) that if H_0 is any self-adjoint operator on a separable Hilbert space, there exist self-adjoint operators V, even of Schmidt class, for which $H_0 + V$ can have a pure point spectrum. (See also Kuroda [1].)

The problem as to whether the continuous spectrum—that is, those points having positive "continuous spectrum multiplicity" (cf. Stone [1], p. 267; Halmos [3], in particular, the remarks on p. 112)—was invariant under even one-dimensional perturbations was open for a long time. The issue was raised by Weyl [2], p. 252, in connection with the operators arising in his theory of singular boundary value problems, and, in a more general setting, by Wintner (cf. Hartman [1], p. 239). For some relevant results in the case of differential operators, see the papers of Aronszajn [1] and Putnam [6]. In the former paper, Aronszajn, using results of Gelfand and Levitan [1] and Krein [2], constructs a one-dimensional boundary value problem with the property that for one boundary condition the spectrum is a pure point spectrum dense on the real axis, while for all other boundary conditions it is purely continuous.

Despite the fact that the continuous spectrum fails to be invariant, even under one-dimensional perturbations, nevertheless, as has been discussed in this chapter, the absolutely continuous spectrum not only remains invariant, but is even unitarily invariant, under a large class of perturbations, namely those of trace class. As mentioned earlier, results of a similar nature go back at least to the work of Friedrichs [1, 2] on certain perturbations of the operator of multiplication by x on $L^2(-1, 1)$.

The theory later evolved into the results of Kato, Rosenblum and others as described above.

Using a method of Friedrichs, Rejto [1] obtains conditions, involving "gentle" perturbations, for the unitary equivalence of the continuous parts of two self-adjoint operators.

Rejto [2] defines in analogous fashion the notion of the part of a self-adjoint operator over a given interval, this being the restriction of the operator to the eigenspace belonging to that interval, that is, to the range of the projection measure associated with the interval by the spectral family of the operator. He then obtains criteria for the unitary equivalence of the parts, as well as the continuous parts, of such operators. The investigations are continued in Rejto [3, 4] wherein are given abstract conditions for the absolute continuity of the part of a self-adjoint operator over an interval and applications of the results to differential operators.

Chapter VI

Laurent and Toeplitz operators, singular integral operators and Jacobi matrices

§ 6.1 Laurent and Toeplitz operators

In this final chapter some applications of the results obtained in the preceding work will be made.

Let $\mathfrak{H}=L^2(0, 2\pi)$ denote the Hilbert space of functions $x=x(t)$,

$$x = x(t) \sim \sum_{-\infty}^{\infty} x_n e^{int}, \quad \sum_{-\infty}^{\infty} |x_n|^2 < \infty, \tag{6.1.1}$$

and let x^+ and x^- be defined by

$$x^+ \sim \sum_{0}^{\infty} x_n e^{int}, \quad x^- \sim \sum_{1}^{\infty} x_{-n} e^{-int}. \tag{6.1.2}$$

The subspaces of elements x^+ and x^- will be denoted respectively by $\mathfrak{H}^+$ and $\mathfrak{H}^-$. If $f=f(t)\in\mathfrak{H}$, so that

$$f = f(t) \sim \sum_{-\infty}^{\infty} f_n e^{int}, \quad \sum_{-\infty}^{\infty} |f_n|^2 < \infty, \tag{6.1.3}$$

then the Laurent operator $L=L_f$ is defined by

$$L_f x = fx, \quad \mathfrak{D}_{L_f} = \{x \in L^2(0, 2\pi) : fx \in L^2(0, 2\pi)\}. \tag{6.1.4}$$

It is clear that L_f is normal, and is self-adjoint if and only if f is real. Furthermore, the spectrum of L_f is the set of values λ with the property that for every $\varepsilon > 0$ the set of values t on $(0, 2\pi)$ for which $|f(t)-\lambda| < \varepsilon$ has positive measure. The point spectrum of L_f consists of those numbers for which $f(t)=\lambda$ holds on a set of positive measure.

If $x, f \in L^2(0, 2\pi)$, then $y=fx \in L(0, 2\pi)$ and y has a representation $y=y^+ + y^-$ where $y^+ \sim \sum_{0}^{\infty} y_n e^{int}$, $y^- \sim \sum_{1}^{\infty} y_{-n} e^{-int}$. In particular, the above holds if $x=x^+$ of (6.1.2). Then the Toeplitz operator $T=T_f$ is defined by

$$T_f x^+ = y^+, \quad \mathfrak{D}_{T_f} = \{x^+ \in \mathfrak{H}^+ : fx^+ = y^+ + y^-, y^+ \in \mathfrak{H}^+\}. \tag{6.1.5}$$

It was shown by Hartman and Wintner [2], p. 878, that if f is real and of class $L^2(0, 2\pi)$ then T_f need not be self-adjoint, but that it is surely self-adjoint if also f is half-bounded, that is, if $f(t) \geqq$ const. $> -\infty$ or $f(t) \leqq$ const. $< \infty$. The condition is not necessary however; Hartman [3], p. 74, Ismagilov [1], p. 465. The determination of the spectrum of T_f, when it is self-adjoint, is considerably more difficult than the corresponding problem for the operator L_f.

Let $f(t)$ be bounded,

$$|f(t)| < \text{const. a.e.}, \qquad (6.1.6)$$

so that both L_f and T_f are bounded. If P is the orthogonal projection of $\mathfrak{H}$ onto $\mathfrak{H}^+$ defined by

$$x^+ = Px, \qquad (6.1.7)$$

with x, x^+ given by (6.1.1) and (6.1.2), it is seen that T_f is given by

$$T_f x^+ = PL_f x^+. \qquad (6.1.8)$$

The operator T_f is said to be analytic if $f \in \mathfrak{H}^+$ and coanalytic if $\bar{f} \in \mathfrak{H}^+$.

Let f satisfy (6.1.6). In terms of ordinary Lebesgue measure on the unit circle, L_f has a (doubly infinite) matrix representation $A = (a_{ij})$ with respect to the orthonormal basis $\{e_n\}$, where $e_n = (2\pi)^{-\frac{1}{2}} e^{int}$ and $n = 0, \pm 1, \ldots$, given by $a_{ij} = (Le_j, e_i) = f_{i-j}$, with $i, j = 0, \pm 1, \ldots$. Similarly, T_f has a (singly infinite) matrix representation given by (f_{i-j}) with $i, j = 0, 1, \ldots$. It can be mentioned that it is possible to start with a corresponding matrix representation $A = (a_{i-j})$ and to show that A is in fact a Laurent or Toeplitz operator; see Brown and Halmos [1].

For use below it is convenient to define the unitary operator U on $\mathfrak{H}$ (two-sided shift) by

$$Ux = e^{it} x, \quad x \in \mathfrak{H}. \qquad (6.1.9)$$

Remark. Self-adjoint Laurent and Toeplitz operators were defined originally in terms of matrices. For early work on the theory of self-adjoint Laurent matrices, see Toeplitz [1, 2], F. Riesz [1], pp. 152–155. The theory of self-adjoint Toeplitz matrices goes back to Toeplitz [3]; the investigation of their spectra was begun by Hartman and Wintner [1,2].

§ 6.2 A spectral inclusion theorem

The next result was proved by Hartman and Wintner [1].

Theorem 6.2.1. *Let f be real and satisfy (6.1.6), so that L_f and T_f are bounded self-adjoint operators. Then the spectrum of L_f is a subset of the spectrum of T_f.*

Proof. Let $\lambda \in \text{sp}(L_f)$. Since $L_{f-\lambda} = L_f - \lambda I$ and $T_{f-\lambda} = T_f - \lambda I$ there is no loss of generality in supposing that $\lambda = 0$. Then for any $\varepsilon > 0$, there exists an $x \in \mathfrak{H}$ $(= L^2(0, 2\pi))$ satisfying $\|x\| = 1$ and $\|L_f x\| < \varepsilon$. Let $x^j = PU^j x (\in \mathfrak{H}^+)$ where P and U are defined by (6.1.7) and (6.1.9). Then

$$x = e^{-ijt}(x^j + \eta^j),$$

where

$$s\text{-}\lim_{j \to \infty} \eta^j = 0.$$

Clearly $\|x^j\| \to \|x\|$ and $\|fx^j\| \to \|fx\|$ and hence

$$\|T_f x^j\| = \|PL_f x^j\| \le \|L_f x^j\| = \|fx^j\| \to \|fx\| = \|L_f x\|.$$

Thus $\|x^j\| > 1 - \varepsilon$ and $\|T_f x^j\| < 2\varepsilon$ for j sufficiently large, and the proof is complete.

§ 6.3 A special Toeplitz matrix

Let the power series $F(z)$, defined by

$$F(z) = \sum_{n=0}^{\infty} f_n z^n, \qquad (6.3.1)$$

be convergent for $|z| < 1$, and put $A = A_F = (a_{ij})$, where $i, j = 0, 1, \ldots$ and $a_{ij} = f_{j-i}$ for $i \le j$ and $a_{ij} = 0$ otherwise. The upper triangular matrix A_F is known to be bounded on l^2 if and only if $|F(z)| < $ const. on $|z| < 1$ (Toeplitz; cf. Hartman and Wintner [2], p. 880) that is, if and only if

$$f^+ \text{ is bounded}. \qquad (6.3.2)$$

If also $G(z) = \sum_{n=0}^{\infty} g_n z^n$ is convergent for $|z| < 1$, then

$$A_{F+G} = A_F + A_G, \quad A_{cF} = cA_F \ (c = \text{const.}), \text{ and } A_1 = I$$

(whether or not $F(z)$ and/or $G(z)$ are bounded in $|z| < 1$). Furthermore, since $F(z)G(z) = \sum_{n=0}^{\infty} h_n z^n$, where $h_n = \sum_{m=0}^{n} f_{n-m} g_m$, converges for $|z| < 1$, it follows that $A_{FG} = A_F A_G$.

The following is due to Wintner [1].

Theorem 6.3.1. *Let $F(z)$ of (6.3.1) be bounded in $|z| < 1$. Then the spectrum of the (bounded) operator A_F is the closure of the range of $F(z)$ on $|z| < 1$.*

Proof. Let $\lambda = F(z_0)$ for some z_0 satisfying $|z_0| < 1$. Since $A_{F-\lambda I} = A_F - \lambda I$ this implies that $(1, z_0, z_0^2, \ldots)$ is an eigenvector of A_F belonging to λ and, in particular, that $\lambda \in \text{sp}(A_F)$. Since $\text{sp}(A_F)$ is a closed set, the closure of the range of $F(z)$ on $|z| < 1$ belongs to $\text{sp}(A_F)$. Next, let μ be any number not in this closure, so that $|\mu - F(z)| > \text{const.} > 0$ for $|z| < 1$. Then $(\mu - F(z))^{-1}$ is a bounded analytic function on $|z| < 1$ and hence

$A_{(F-\mu)^{-1}}$ is the (bounded) inverse of $A_F - \mu I$ ($= A_{F-\mu I}$). This completes the proof of Theorem 6.3.1.

Suppose that f satisfies (6.3.2), so that A_F is bounded. A straightforward verification shows that

$$A_F A_F^* - A_F^* A_F = B^* B, \quad B = (f_{i+j+1}), \quad i, j = 0, 1, \ldots, \qquad (6.3.3)$$

and hence A_F is semi-normal. The question was raised in § 3.5 as to whether the inequality (3.5.17) is valid for all semi-normal operators. It will be shown that the relation is true for A_F at least if the mapping

$$z \to F(z) \text{ on } |z| < 1 \text{ is one-to-one} . \qquad (6.3.4)$$

In order to see this note that in view of Theorem 6.3.1, the (two-dimensional) measure of sp(T) is not greater than the double integral of $|F'(z)|^2$ taken over the disk $|z| \leq 1$, that is,

$$\operatorname{meas}_2 (\operatorname{sp}(A_F)) \leq \pi \sum_{k=1}^{\infty} k |f_k|^2 ,$$

with equality certainly holding if (6.3.4) is assumed. On the other hand, it is easily shown that

$$\| B^* B \| \leq \sum_{k=1}^{\infty} k |f_k|^2$$

(cf. Putnam [32]). Consequently, one has (3.5.17), that is,

$$\pi \| B^* B \| \leq \operatorname{meas}_2 (\operatorname{sp}(A_F)), \qquad (6.3.5)$$

at least if (6.3.4) holds. (Note that (3.5.16) is valid for any semi-normal operator, and, in particular, holds in the present instance. Consequently, relation (6.3.5) surely holds whenever sp(A_F) is convex or, more generally, satisfies the conditions of Corollary 2 of § 3.5.)

§ 6.4 Spectra of self-adjoint Toeplitz operators

The following is due to Hartman and Wintner [2].

Theorem 6.4.1. Let $f(t)$ of (6.1.3) be bounded, real and measurable on $(0, 2\pi)$ and put $m = \operatorname{ess} \inf f$, $M = \operatorname{ess} \sup f$ on $(0, 2\pi)$. Then

$$\operatorname{sp}(T_f) = [m, M] . \qquad (6.4.1)$$

Proof. A proof will be given only for the case that f satisfies (6.3.2). (Actually the theorem holds without this hypothesis however. In fact, an analogous assertion holds whenever T_f is self-adjoint though possibly not bounded. For a general proof, see Hartman and Wintner [2].) The reason for assuming (6.3.2) in the present argument is to ensure an operator boundedness condition needed in order to apply a result (Theorem 3.4.1) on semi-normal operators.

Since $T_{f+\lambda} = T_f + \lambda I$ there is no loss of generality in supposing that $f_0 = 0$, thus

$$f_{-n} = \bar{f}_n, \quad f_0 = 0 . \tag{6.4.2}$$

If $G(z) = \sum_{n=1}^{\infty} \bar{f}_n z^n$, then (6.3.2) implies that A_G is bounded. It is easily verified that T_{f+} corresponds to A_G^* and that (cf. (6.3.3))

$$T_{f+}^* T_{f+} - T_{f+} T_{f+}^* = C^* C , \tag{6.4.3}$$

where C on $\mathfrak{H}^+$ corresponds to the matrix $(\bar{f}_{i+j+1})$ on l^2, so that T_{f+} is hyponormal. Since $\operatorname{sp}(T_{f+}) = \overline{\operatorname{sp}(A_G)}$ is the closure of the range of $\bar{G}(z)$ on $|z| < 1$ (see Theorem 6.3.1), its projections on the real and imaginary axes are closed intervals. By Theorem 3.4.1, the projection on the real axis is the spectrum of $\frac{1}{2}(T_{f+} + T_{f+}^*)$. In virtue of (6.4.2), $f = f^+ + \bar{f}^+$ and so $T_f = T_{f+} + T_{f+}^*$, hence $\operatorname{sp}(T_f)$ is an interval. There remains only to show that the end-points of this interval are m and M. But the latter numbers are clearly the minimum and maximum points of $\operatorname{sp}(L_f)$ and so, by Theorem 6.2.1, m and M belong to $\operatorname{sp}(T_f)$. It is clear from the definition (6.1.5) however that $\operatorname{sp}(T_f) \subset [m, M]$ and the proof is now complete.

§ 6.5 Two lemmas

The following lemmas were proved in Douglas and Pearcy [1]. (Recall that $[\mathfrak{M}]$ denotes the closure of a linear manifold $\mathfrak{M}$.)

Lemma 6.5.1. *If* $L = L_f$ *is the Laurent operator belonging to a bounded measurable function* $f = f(t)$ *of (6.1.3) and if* $[L\mathfrak{H}] \neq \mathfrak{H}$ *then* $[L\mathfrak{H}] = [L\mathfrak{H}^+] = [L\mathfrak{H}^-]$.

Proof. That $[L\mathfrak{H}^+] \subset [L\mathfrak{H}]$ is obvious. It will be shown that if $x_0 \in [L\mathfrak{H}]$ and $x_0 \perp [L\mathfrak{H}^+]$, then $x_0 = 0$. To this end, let E denote the projection defined by $E : \mathfrak{H} \to [L\mathfrak{H}]$ and define the sets $R = \{t \in [0, 2\pi], f(t) \neq 0\}$, $S = [0, 2\pi] - R$. (Clearly $L^2(R) = E\mathfrak{H}$.) Since $L = LE(=EL)$, then $(I - E)L^* = 0$ and hence

$$(I - E) L^* x_0 = 0 . \tag{6.5.1}$$

But if y^+ is any element in $\mathfrak{H}^+$ then, since $x_0 \perp [L\mathfrak{H}^+]$,

$$(L^* x_0, y^+) = (x_0, Ly^+) = 0, \text{ and so } L^* x_0 \in \mathfrak{H}^- .$$

But $I - E = g_S$, the multiplication operator corresponding to the characteristic function of the set S. Since $[L\mathfrak{H}] \neq \mathfrak{H}$, then $g_S \neq 0$ on a set of positive measure and, since $L^* x_0 \in \mathfrak{H}^-$, it follows from (6.5.1) and the F. and M. Riesz theorem [1] that $L^* x_0 = 0$. Thus $x_0 \perp [L\mathfrak{H}]$ and, since $x_0 \in [L\mathfrak{H}]$, $x_0 = 0$. Consequently $[L\mathfrak{H}] = [L\mathfrak{H}^+]$. The proof of $[L\mathfrak{H}] = [L\mathfrak{H}^-]$ is similar and the proof is now complete.

Lemma 6.5.2. *Let* $L = L_f$ *be the Laurent operator belonging to a* *bounded measurable function $f=f(t)$ of* (6.1.3), *let P be defined by* (6.1.7), *and let Q denote a projection on $\mathfrak{H}$ satisfying the conditions*

$$Q \neq 0, \ Q \neq I \ \text{and either} \ P \leq Q \ \text{or} \ Q \leq P \tag{6.5.2}$$

and

$$LQ = QL. \tag{6.5.3}$$

Then $L = \lambda I$ for some number λ.

Proof. Suppose first that $P \leq Q$. If the Lemma is false there would exist a spectral projection E belonging to L satisfying $E \neq 0, I$. If $\mathfrak{M} = Q\mathfrak{H}$ then $\mathfrak{H}^+ = P\mathfrak{H} \subset \mathfrak{M}$. By (6.5.3), $EQ = QE$ and hence $E\mathfrak{M} \subset \mathfrak{M}$. If now E is identified with L of Lemma 6.5.1, it follows that $E\mathfrak{H} = [E\mathfrak{H}^+]$. But $[E\mathfrak{H}^+] \subset [E\mathfrak{M}] \subset \mathfrak{M}$ and so $E\mathfrak{H} \subset \mathfrak{M}$. A similar argument with E replaced by $I - E$ shows that $(I - E)\mathfrak{H} \subset \mathfrak{M}$ and hence $\mathfrak{H} \subset \mathfrak{M}$. Thus $Q = I$, a contradiction.

The proof in case $Q \leq P$ is similar. For $I - P \leq I - Q$ and so $\mathfrak{N} \equiv (I - Q)\mathfrak{H} \supset (I - P)\mathfrak{H} = \mathfrak{H}^-$. An application of Lemma 6.5.1 implies that $E\mathfrak{H} = [E\mathfrak{H}^-] \subset [E\mathfrak{N}] \subset \mathfrak{N}$ and so $E\mathfrak{H} \subset \mathfrak{N}$. Similarly $(I - E)\mathfrak{H} \subset \mathfrak{N}$, hence $\mathfrak{H} \subset \mathfrak{N}$, and so $Q = 0$, a contradiction.

Remark. Concerning the F. and M. Riesz theorem see also § 2.8.

§ 6.6 Analytic and coanalytic Toeplitz operators

Theorem 6.6.1. *Let f of* (6.1.3) *be bounded and measurable. Suppose that T_f is either an analytic $(f = f^+)$ or coanalytic $(\bar{f} = \bar{f}^+)$ operator on $\mathfrak{H}^+$ and that $\mathfrak{M}$ is a non-trivial subspace of $\mathfrak{H}^+$ reducing T_f on which T_f is normal. Then $f \equiv$ const.*

Proof. The proof depends on the lemmas of § 6.5 and is due to R. G. Douglas. It can be supposed that T_f is analytic (since, if T_f is coanalytic, then $T_f^* = T_{\bar{f}}$ is analytic). If $x \in \mathfrak{H}^+$, then $T_f x = fx = L_f x$ and so $\mathfrak{M}$ is invariant under L_f. Since the restriction of L_f to $\mathfrak{M}$ is normal, then $\mathfrak{M}$ reduces L_f on $\mathfrak{H}$; cf., e.g., Berberian [2], p. 160. Let Q denote the projection $Q : \mathfrak{H} \to \mathfrak{M}$. Then, since $\mathfrak{M} \subset \mathfrak{H}^+$, $Q \leq P$. Moreover $Q \neq 0$, I and $L_f Q = Q L_f$. Hence, by Lemma 6.5.2, L_f is a scalar, so that $f \equiv$ const.

§ 6.7 Absolute continuity of Toeplitz operators

Theorem 6.7.1. *Let $f(t)$ be bounded, real and measurable and suppose that $f(t) \not\equiv$ const. In addition, suppose that* (6.3.2) *holds. Then $T = T_f$ is absolutely continuous, that is $\mathfrak{H}_a(T_f) = \mathfrak{H}^+$.*

Proof. Clearly it can be supposed that $f_0 = 0$, so that (6.4.2) holds. Since $T_f = 2 \operatorname{Re}(T_{f+})$, it follows from (6.4.3) and Theorem 3.2.1 that $\mathfrak{H}_a(T_f)$ contains the least space Ω reducing T_{f+} and containing the range of C, and that T_{f+} is normal on $\Omega^\perp$. But if $\Omega^\perp \neq 0$, it follows from Theorem 6.6.1 that $f^+ \equiv$ const. $(= 0)$ and hence $f \equiv 0$, a contradiction. Hence $\Omega^\perp = 0$, that is $\Omega = \mathfrak{H}^+$, and the proof is complete.

Special cases of Theorem 6.7.1 were obtained in Putnam [13,16,18].

In case it is assumed only that f is real, of class $L^2(0, 2\pi)$ and half-bounded (hence T_f is self-adjoint) it was shown by Rosenblum [6], using the representation of Aronszajn and Donoghue [1], that T_f is absolutely continuous. Further, Ismagilov [1] (cf. also Rosenblum [8]) asserts that the conclusion of Theorem 6.7.1 holds under the minimum hypothesis, that is, T_f must be absolutely continuous whenever it is self-adjoint.

An idea of Ismagilov's argument is the following. Let $A = \int \lambda \, dE_\lambda$ be a self-adjoint operator on a Hilbert space $\mathfrak{H}$ and suppose that A has a simple spectrum. Let $\phi \in \mathfrak{D}_A$ be a generating vector, so that $\mathfrak{H}$ is the closure of the linear manifold determined by $\{E_\lambda \phi\}$. If P_ϕ denotes the orthogonal projection of $\mathfrak{H}$ onto $\mathfrak{H}_\phi = \mathfrak{H} \ominus \{\phi\}$ define the self-adjoint operator A_ϕ in $\mathfrak{H}_\phi$ by $A_\phi x = P_\phi A x$ for $x \in \mathfrak{D}_A \cap \mathfrak{H}_\phi$. Then the absolutely continuous parts of A and A_ϕ are unitarily equivalent, while the singular parts of the corresponding spectral families have (some pair of) disjoint supports. (This result is analogous to a theorem of Aronszajn [1], p. 602, in connection with boundary value problems.) It follows in particular that whenever A and A_ϕ are unitarily equivalent then their singular parts must be absent. This last fact is exploited to prove that any self-adjoint Toeplitz operator must be absolutely continuous.

It can be mentioned that Ismagilov asserts a complete spectral analysis for T_f. The following theorem is used: if A, B and $A + B$ are self-adjoint operators with product AB of trace class, then the absolutely continuous part of $A + B$ is unitarily equivalent to the direct sum of the absolutely continuous parts of A and B.

Also, Ismagilov refines the self-adjointness conditions of Hartman and Wintner [2], p. 878, by stating that T_f must be self-adjoint if $f(t)$ is half-bounded in some neighborhood of every point of $[0, 2\pi]$. See Hartman [3] for other criteria for self-adjointness.

§ 6.8 Spectral resolutions for certain Toeplitz operators

Let $f(t)$ of (6.1.3) be real and even, and suppose that $f_0 = 0$, so that

$$f(t) \sim 2 \sum_{n=1}^{\infty} f_n \cos nt, \tag{6.8.1}$$

where the f_n are real. Let J denote the matrix belonging to the quadratic form $2 \sum_{n=0}^{\infty} x_n x_{n+1}$, so that $J = 2(V + V^*)$, where V is the unilateral shift on l^2. Then the differential $d\rho_{ij}(t)$, $0 \leq t \leq \pi$, of the spectral matrix of J is given by

$$d\rho_{ij}(t) = 2\pi^{-1}[\sin(i+1)t][\sin(j+1)t] \, dt \quad (i, j = 0, 1, \ldots) \, ;$$

cf. Hilbert [1], p. 155, Hellinger [1], pp. 148 ff. A straightforward calculation shows that

$$A_f = F + K, \quad \text{where } A_f = (f_{i-j}), \tag{6.8.2}$$

and

$$F = \left(\int_0^\pi f(t)\,\mathrm{d}\rho_{ij}(t) \right) \quad \text{and} \quad K = (f_{i+j+2}). \tag{6.8.3}$$

Theorem 6.8.1. *Suppose that $f(t)$ of (6.1.3) is real, even, absolutely continuous, and that its derivative has an absolutely convergent Fourier series, thus*

$$f_{-n} = \bar{f}_n = f_n, \quad \sum_{n=1}^\infty n|f_n| < \infty \tag{6.8.4}$$

Then A_f and the absolutely continuous part of F are unitarily equivalent.

Proof. If $\{e_n\}$, $n = 0, 1, \ldots$, denotes the complete orthonormal system of vectors $(1, 0, 0, \ldots)$, $(0, 1, 0, \ldots)$, $\ldots$ on l^2, then

$$\mathrm{tr}|K| = \Sigma(|K|e_n, e_n) \leq \Sigma \||K|e_n\| = \Sigma \|Ke_n\|$$

$$= \sum_{n=0}^\infty \left(\sum_{m=0}^\infty |f_{n+m+2}|^2 \right)^{\frac{1}{2}} \leq \sum_{n=0}^\infty \sum_{m=0}^\infty |f_{n+m+2}| \leq \sum_{n=1}^\infty n|f_{n+1}| < \infty .$$

Hence the Hankel matrix K is of trace class. But A_f is absolutely continuous by Theorem 6.7.1 and so the assertion follows from the Rosenblum-Kato theorem (Theorem 5.5.1). See Putnam [16,18], Rosenblum [6,7,8].

For some of the theory of not necessarily self-adjoint Toeplitz operators, including the role of the work of Beurling [1] and Halmos [8] and also for further references on the subject, consult Hartman [3]. See also Devinatz [1].

Concerning the connection between self-adjoint Toeplitz operators and Hankel matrices see Putnam [13, 16, 28], Rosenblum [6].

Further material and references concerning Toeplitz operators can be found in Grenander and Szegö [1], Halmos [9], Widom [2] and in other papers cited in this and preceding sections.

§ 6.9 Some results for unbounded operators

It has been indicated that some of the above results can be generalized to unbounded Laurent or Toeplitz operators. Thus, this was the situation in connection with Theorem 6.7.1. The proof given above of Theorem 6.7.1 as stated, in which (6.3.2) was assumed, was based on results on commutators. Whether these methods can be used to obtain the absolute continuity of arbitrary self-adjoint Toeplitz operators (Ismagilov [1]; Rosenblum [8]) or even of half-bounded ones (Rosenblum [6]) will remain undecided.

In this connection however the following remarks may be of some interest. First it can be observed that for arbitrary $f \in L^2(0, 2\pi)$, T_f is defined but T_f^* is in general not $T_{\bar{f}}$ (cf. Hartman [3], p. 63). On the other hand, if $T = T_f$ and $S = T_{\bar{f}}$ and if $f = f^+$ then it is easily shown that

$$ST - TS = D, \qquad D \geq 0, \tag{6.9.1}$$

holds on $\Omega = \mathfrak{D}_{ST} \cap \mathfrak{D}_{TS} (\subset \mathfrak{H}^+)$. In fact, if $x^+ \in \Omega$ then $STx^+ = P(\bar{f}fx^+)$ and hence $(STx^+, x^+) = \|fx^+\|^2$. Also $TSx^+ = f(P\bar{f}x^+)$ and hence $(TSx^+, x^+) = \|P\bar{f}x^+\|^2 \leq \|\bar{f}x^+\|^2 = \|fx^+\|^2$. Hence (6.9.1) holds. It is clear also that $(T+S)(T-S) - (T-S)(T+S) = 2D$ holds on Ω. Consequently $AJ - JA = -iC$, where $C \geq 0$ and $A = T_{f+\bar{f}}$, $J = T_{(f-\bar{f})/i}$, holds on Ω. In case both operators A and J are self-adjoint, the results of § 2.13 are applicable. Such an application will be given in the next section.

§ 6.10 Hilbert matrix

The Hilbert matrix A defined on l^2 is given by $A = (a_{jk})$ with $a_{jj} = 0$ and $a_{jk} = |j-k|^{-1}$ for $j \neq k$, where $j, k = 0, 1, \ldots$. It is easily verified that the matrix equation

$$AJ - JA = -2iH^2 \tag{6.10.1}$$

holds, where $J = (b_{jk})$ with $b_{jj} = 0$ and $b_{jk} = i(k-j)^{-1}$ for $j \neq k$ and $H = ((j+k+1)^{-1})$. It is known that both J and H are bounded (Hilbert); cf. Hardy, Littlewood and Pólya [1], pp. 212–213, 223. Furthermore, the spectrum of H is the interval $[0, \pi]$ and is purely continuous (Magnus [1]). In addition A is known to be unbounded; cf. Hardy, Littlewood and Pólya [1], p. 214. Since A is the Toeplitz operator $A = A_f$ belonging to

$$f(t) = -2 \log(2|\sin \tfrac{1}{2}t|) \sim 2 \sum_{n=1}^{\infty} n^{-1} \cos nt \text{ on } L^2(0, 2\pi)$$

and, since $f(t)$ is half-bounded, A_f is self-adjoint (and half-bounded); Hartman and Wintner [2], p. 878. (Another proof of the self-adjointness of A can be deduced from the fact that H is bounded; cf. Putnam [16], p. 844.) Furthermore it is easy to show (Putnam [16], loc. cit.) that (6.10.1) is valid as an operator equation on the set $\mathfrak{D}_A$. Clearly the set Ω_λ of Theorem 2.13.2 is $\mathfrak{D}_A$ for all $\lambda \neq 0$ and the absolute continuity of A as well as of J then follows from that theorem.

For other results on the Hilbert matrix, see also Hardy, Littlewood and Pólya [1], Hill [1], Kato [5], Magnus [1], Rosenblum [3, 4].

§ 6.11 Singular integral operators

Let E denote a set of real numbers of positive (possibly infinite) measure and let $h(t)$ and $\phi(t)$ be functions defined on E satisfying

$h(t)$ real, $\phi(t)$ complex, h and ϕ measurable
and essentially bounded on E. (6.11.1)

Let $A = A(h, \phi, E)$ be defined by

$$Ax = A(h, \phi, E)x = h(t)x(t) + (i\pi)^{-1} \int_E \bar{\phi}(s)\phi(t)(s-t)^{-1}x(s)\,ds,$$
(6.11.2)

where the integral is a Cauchy principal value at $s = t$, that is,

$$\int_E = \lim_{\varepsilon \to 0+} \int_{E_{\varepsilon t}} \quad \text{where} \quad E_{\varepsilon t} = E \cap \{s : |s-t| > \varepsilon\}.$$

The operator A is a bounded self-adjoint transformation of $L^2(E)$ into itself; see, e.g., Muskhelishvili [1], Schwartz [1]. In case $h(t) \equiv 0$ and $\phi(t) \equiv 1$, the transformation of (6.11.2) becomes

$$A(0, 1, E)x = (i\pi)^{-1} \int_E (s-t)^{-1}x(s)\,ds,$$
(6.11.3)

which, in case $E = (-\infty, \infty)$, becomes the Hilbert transformation

$$Hx = A(0, 1, (-\infty, \infty))x = (i\pi)^{-1} \int_{-\infty}^{\infty} (s-t)^{-1}x(s)\,ds.$$
(6.11.4)

The operator H is unitary and its spectrum consists of ± 1, each of infinite multiplicity; see Titchmarsh [3]. As to the role of the Hilbert transform in the dispersion relations of physics, see Hilgevoord [1].

Concerning singular integral operators similar to (6.11.2) or (6.17.2) below, see also the remarks at the end of § 1.7. A list of references can be found in the bibliography of Muskhelishvili [1]. Some more recent work is that of Calderon and Zygmund [1], Freeman [1], Koppelman [1, 2], Koppelman and Pincus [1, 2], Pincus [1,2], Putnam [29, 31], Schwartz [1, 2, 3, 4], Shinbrot [1], Widom [1], Xa Dao-xeng [1]. See also the recent survey paper of Calderon [1] wherein can be found numerous references.

Some properties of the operator $A(h, \phi, E)$ will be derived in the sections below. The methods used will involve applications of general results on commutators in Hilbert space which were obtained earlier. The treatment is based largely on that in Putnam [29, 31].

§ 6.12 $A(h, \phi, E)$ with E bounded

Theorem 6.12.1. *Suppose that $E = (a, b)$ is a finite interval and that h, ϕ and $A = A(h, \phi, (a, b))$ are defined by (6.11.1) and (6.11.2). Then*

$$\int_a^b |\phi(t)|^2\,dt \leq \tfrac{1}{2}(b-a) \text{ meas sp}(A).$$
(6.12.1)

Proof. If B is the multiplication operator on $L^2(a, b)$ defined by

$$Bx = tx(t), \tag{6.12.2}$$

then an easy calculation shows that

$$(AB - BA)x = (i\pi)^{-1} \int_a^b \phi(t)\,\overline{\phi}(s)\,x(s)\,ds \equiv iCx; \tag{6.12.3}$$

cf. Friedrichs [1], p. 252, [2], p. 369. Since $Cx = -\pi^{-1}\phi(x, \phi)$, then $(Cx, x) = -\pi^{-1}|(x, \phi)|^2 \leq 0$. In addition, $\|C\| = \pi^{-1}\|\phi\|^2$ and max $\mathrm{sp}(B) - \min\mathrm{sp}(B) = b - a$. The assertion of the theorem now follows from Theorem 2.2.1'.

It can be noted that, if $\phi \equiv 1$ and $a = -b$, the above example proves the optimality of the constant $2/\pi$ in the inequality of (2.1.2). For, in this case, $\|B\| = b$, $\|A\| = 1$ (see (6.13.1) below) and hence

$$0 < \|C\| = (2/\pi)\|A\|\,\|B\|.$$

Corollary. *Let E be a bounded set of positive measure and let h, ϕ and $A = A(h, \phi, E)$ be defined by (6.11.1) and (6.11.2). Then*

$$\int_E |\phi|^2 \, dt \leq \tfrac{1}{2}(\sup E - \inf E)\,\mathrm{meas}\,\mathrm{sp}(A). \tag{6.12.4}$$

Proof. Let $[a, b]$ contain E and let the domain of definition of h and ϕ be extended to $[a, b]$ by putting $h = \phi = 0$ on $[a, b] - E$. Define A_{ab} on $L^2(a, b)$ by

$$A_{ab}x = (i\pi)^{-1} \int_a^b \overline{\phi}(s)\,\phi(t)(s - t)^{-1} x(s)\,ds.$$

Then $L^2(E)$ reduces A_{ab} and A_{ab} is the direct sum of $A(h, \phi, E)$ on $L^2(E)$ and the zero operator on $L^2((a, b) - E)$. Since $\mathrm{meas}\,\mathrm{sp}(A_{ab}) = \mathrm{meas}\,\mathrm{sp}(A)$, an application of Theorem 6.12.1 to A_{ab}, when $a = \inf E$ and $b = \sup E$, proves the Corollary.

§ 6.13 The norm of $A(0, \phi, E)$

Theorem 6.13.1. *Let $\phi(t)$ be defined as in (6.11.1) and suppose that $h(t) \equiv 0$. Then*

$$\|A(0, \phi, E)\| = \operatorname*{ess\,sup}_E |\phi(t)|^2 \tag{6.13.1}$$

and

$$\pm\|A(0, \phi, E)\| \text{ are in the essential spectrum of } A(0, \phi, E). \tag{6.13.2}$$

Proof. Let the domain of definition of ϕ be extended to $(-\infty, \infty)$

by putting $\phi = 0$ on the complement of E and then define A' on $L^2(-\infty,\infty)$ by

$$A'x = (i\pi)^{-1} \int_{-\infty}^{\infty} \bar{\phi}(s)\,\phi(t)(s-t)^{-1}x(s)\,\mathrm{d}s .$$

It is clear that $L^2(E)$ reduces A', that A' is the direct sum of $A = A(0, \phi, E)$ on $L^2(E)$ and the zero operator on $L^2((-\infty, \infty) - E)$, and that $\|A\| = \|A'\|$. If $y = \bar{\phi}x$ and $v = \operatorname*{ess\,sup}_{E}|\phi|^2$, it is clear that $\|A'x\|^2 \leq v\|Hy\|^2$, where H is defined by (6.11.4). Since $\|H\| = 1$, then $\|A'\| \leq v$.

Next, let (a, b) be any finite interval and let $P = P_{ab}$ denote the projection of $L^2(-\infty, \infty)$ onto $L^2(a, b)$. An application of Theorem 6.12.1 yields, for arbitrary $a < b$,

$$2(b-a)^{-1} \int_{a}^{b} |\phi|^2\,\mathrm{d}t \leq \operatorname{meas} \operatorname{sp}(PA'P) \leq 2\|PA'P\| \leq 2\|A'\|$$

for all $P = P_{ab}$. On letting $b \to a + 0$, it follows that $|\phi(a)|^2 \leq \|A'\|$ for almost all a, hence $v \leq \|A'\|$ and the proof of (6.13.1) is complete.

In order to prove (6.13.2), note from the preceding argument that, for every $\varepsilon > 0$, $P = P_{ab\varepsilon}$ can be chosen so that $\operatorname{meas} \operatorname{sp}(PA'P) > 2\|A\| - \varepsilon$. Hence there exist α, β in $\operatorname{sp}(PA'P)$ satisfying $\alpha < -\|A\| + \varepsilon$ and $\beta > \|A\| - \varepsilon$. Consequently there exist unit vectors $x, y \in L^2(a, b)$ for which $(PA'Px, x) = (A'x, x) < -\|A\| + \varepsilon$ and $(PA'Py, y) = (A'y, y) > \|A\| - \varepsilon$. If now $\varepsilon_1, \varepsilon_2, \ldots$ are chosen so that $\varepsilon_n \to 0$ and the intervals (a_n, b_n) are chosen so that $b_n - a_n \to 0$, it is clear that the corresponding sequences $\{x_n\}, \{y_n\}$ of unit vectors in $L^2(a_n, b_n)$, now regarded in the natural way as a subspace of $L^2(-\infty, \infty)$, converge weakly to 0 in $L^2(-\infty, \infty)$ and satisfy $(A'x_n, x_n) \to -\|A\|$ and $(A'y_n, y_n) \to \|A\|$. Thus $((A' + \|A\|I)x_n, x_n) \to 0$ and hence, since $A' + \|A\|I \geq 0$, there exists the strong limit $(A' + \|A\|I)x_n \to 0$. Similarly, $(A' - \|A\|I)y_n \to 0$. That $\pm\|A\|$ belong to the essential spectrum of A', hence also of A, now follows from Weyl's theorem (Weyl [1]; cf., Riesz and Sz.-Nagy [1], p. 364), and the proof is complete.

Note that if $E = (-\infty, \infty)$ and $\phi(t) \equiv 1$, then $A = H$, the Hilbert transformation (6.11.4), and the essential spectrum in this case coincides with the spectrum, namely the two points $\pm 1 (= \pm\|H\|)$.

§ 6.14 An estimate of meas sp($A(h, \phi, E)$)

Theorem 6.14.1. *Let* h, ϕ *satisfy* (6.11.1) *and let* $t = c$ *have the property that for some* $\varepsilon > 0$,

$$E \cap (c, c+\varepsilon) \text{ has measure } 0. \tag{6.14.1}$$

If $A = A(h, \phi, E)$ *then*

$$\text{meas sp}(A) \geq 2 \lim_{\mu \to c+0} \sup (\mu - c) \int_E |\phi(t)|^2 (t - \mu)^{-2} \, dt. \quad (6.14.2)$$

Similarly, if for some $\varepsilon > 0$,

$$E \cap (c - \varepsilon, c) \text{ has measure } 0 \quad (6.14.3)$$

then

$$\text{meas sp}(A) \geq 2 \lim_{\mu \to c-0} \sup (c - \mu) \int_E |\phi(t)|^2 (t - \mu)^{-2} \, dt. \quad (6.14.4)$$

Proof. Since the proofs of (6.14.2) and (6.14.4) are similar it is sufficient to prove (6.14.2) only, assuming (6.14.1). Let $\mu \in (c, c + \varepsilon)$ and define the bounded self-adjoint operator B_μ on $L^2(E)$ by

$$B_\mu x = (t - \mu)^{-1} x(t), \quad x \in L^2(E). \quad (6.14.5)$$

A calculation similar to that of (6.12.3) shows that

$$AB_\mu - B_\mu A = iC_\mu, \quad (6.14.6)$$

where

$$C_\mu x = \pi^{-1} \psi(x, \psi), \quad \psi(t) = (t - \mu)^{-1} \phi(t) \quad (\in L^2(E)). \quad (6.14.7)$$

Since $(C_\mu x, x) = \pi^{-1} |(x, \psi)|^2$ then $C_\mu \geq 0$ and $\|C_\mu\| = \pi^{-1} \|\psi\|^2$ and hence by Theorem 2.2.1′,

$$\|\psi\|^2 \leq \tfrac{1}{2}(M - m) \text{ meas sp}(A), \quad (6.14.8)$$

where M, m denote the maximum and minimum of sp(B_μ). By (6.14.1), it is clear that $M - m \leq (\mu - c)^{-1} - (\mu - (c + \varepsilon))^{-1}$ and (6.14.2) follows from (6.14.8).

Remark. In case the set E is bounded, the operator B of (6.12.2) can be used rather than B_μ. For in this case B is bounded and $AB - BA = iC$ with $C \leq 0$ (cf. (6.12.3)).

Also it can be noted that another proof of Theorem 6.12.1 follows from (6.14.8) if μ is chosen so that $\mu > b$. It need only be observed that

$$M - m = (\mu - b)^{-1} - (\mu - a)^{-1} = (b - a)(\mu - b)^{-1}(\mu - a)^{-1}$$

and that

$$(M - m)^{-1} \|\psi\|^2 \to (b - a)^{-1} \|\phi\|^2 \text{ as } \mu \to \infty.$$

Corollary 1. *Let h, ϕ be defined as in Theorem 6.14.1 and suppose that (6.14.1) holds. Then*

$$\text{meas sp}(A) \geq 2 \lim_{\delta \to 0+} \inf \delta^{-1} \int_{E \cap (c - \delta, c)} |\phi|^2 \, dt. \quad (6.14.9)$$

Similarly, if instead of (6.14.1), relation (6.14.3) is assumed then

$$\text{meas sp}(A) \geq 2 \lim_{\delta \to 0+} \inf \delta^{-1} \int_{E \cap (c, c + \delta)} |\phi|^2 \, dt. \quad (6.14.10)$$

Proof. It is enough to show that (6.14.3) implies (6.14.10). If ϕ is put equal to 0 on the complement of E and if

$$\Phi(t) = \int_c^t |\phi(u)|^2 \, du \, ,$$

then the right side of (6.14.4) is unchanged if the integration domain E is replaced by (c, ∞). Since $\Phi(t) = O(t)$ for large t, an integration by parts shows that

$$\int_c^\infty |\phi(t)|^2 (t-\mu)^{-2} \, dt = 2 \int_c^\infty \Phi(t)(t-\mu)^{-3} \, dt \, ,$$

where $c - \varepsilon < \mu < c$. If α denotes the "lim inf" expression of (6.14.10) then, for any $\beta > 0$, $\Phi(t) \geqq (\alpha - \beta)(t - c)$ for $t - c$ sufficiently small and positive. Replacing $t - c$ by $(t - \mu) + (\mu - c)$ and then estimating the last integral lead to (6.14.10) by virtue of (6.14.4).

Corollary 2. *Let* $E = \bigcup_{n=1}^N [a_n, b_n]$ *be a finite or denumerably infinite union of finite, pairwise disjoint, closed intervals and, if* $N = \infty$, *suppose that* $|a_n| \to \infty$ *as* $n \to \infty$. *Suppose, in addition, that* $|\phi(t)|$ *is monotone on each interval* $[a_n, b_n]$. *Then, if* h *and* ϕ *satisfy* (6.11.1),

$$\operatorname{meas} \operatorname{sp}(A(h, \phi, E)) \geqq 2v \, , \tag{6.14.11}$$

and, if also $h \equiv 0$,

$$\operatorname{sp}(A(0, \phi, E)) = [-v, v] \, , \tag{6.14.12}$$

where $v = \operatorname*{ess\,sup}_E |\phi|^2$.

Proof. The hypothesis implies the existence of points c_n, where each c_n is an end-point of some interval $[a_m, b_m]$ of E, with the property that $v_n = \lim_{t \to c_n} |\phi|^2 \to v$ as $n \to \infty$. Clearly each c_n satisfies either condition (6.14.1) or (6.14.3), and relation (6.14.11) follows from Corollary 1 above. Since, for any self-adjoint operator T, $\operatorname{meas} \operatorname{sp}(T) \leqq 2\|T\|$, the assertion (6.14.12) now follows from Theorem 6.13.1.

§ 6.15 Remarks

It is apparently not known whether (6.14.11) or (6.14.12) holds whenever $E \neq (-\infty, \infty)$, that is, whenever

$$(-\infty, \infty) - E \text{ has positive measure.} \tag{6.15.1}$$

However the validity of (6.14.12) has been established in certain special cases.

Widom [1], p. 152, has shown that (6.14.12) holds whenever

$$|\phi(t)| \equiv 1 \text{ on } E \tag{6.15.2}$$

(hence $v=1$) for any set E satisfying (6.15.1). Widom also considers the operator $A(0, 1, E)$ for general sets E and spaces $L^p(E)$ with $p \geq 1$. (Incidentally, it can be noted that when ϕ satisfies (6.15.2) the operator $A(0, \phi, E)$ is unitarily equivalent to $A(0, 1, E)$. In fact, for any measurable $\phi = \phi(t)$, if U denotes the unitary multiplication operator of multiplication by ϕ on $L^2(E)$, it is clear that $A(0, \phi, E) = UA(0, 1, E)U^*$.

The validity of (6.14.12) in case

$$E \text{ is an interval} \qquad (6.15.3)$$

was established by Koppelman [1], p. 61, at least if $\phi(t)$ is real, positive and has a continuous first derivative possessing only isolated zeros. Moreover, it follows from recent work of Pincus [2] that (6.14.12) is generally true when (6.15.3) holds and ϕ satisfies only (6.11.1).

§ 6.16 Absolute continuity

Theorem 6.16.1. *Let h and ϕ be defined as in* (6.11.1) *and suppose that*

$$E \cap I \text{ is of measure } 0 \text{ for some interval } I \qquad (6.16.1)$$

and that

$$\phi(t) \neq 0 \text{ on } E \text{ a.e.} \qquad (6.16.2)$$

Then $A = A(h, \phi, E)$ of (6.11.2) *is absolutely continuous.*

Proof. Let μ belong to the interior of any interval I satisfying (6.16.1). Define B_μ by (6.14.5) so that (6.14.6) and (6.14.7) hold. It follows from Theorem 2.2.4 that $\mathfrak{H}_a(A)$ contains the least subspace $\mathfrak{L}$ of $\mathfrak{H} = L^2(E)$ reducing both A and B_μ and containing the range of C_μ. Clearly $\mathfrak{L}$ contains the closed linear manifold spanned by the elements $\xi_n = (t-\mu)^{-n}\psi$, $n = 0, 1, 2, \ldots$. But if $x \in L^2(E)$ and $(x, \xi_n) = 0$ for $n = 0, 1, 2, \ldots$, that is,

$$\int_E x(t) \, \overline{\psi}(t)(t-\mu)^{-n} dt = 0, \qquad (6.16.3)$$

then the change of variable $t' = (t-\mu)^{-1}$ in the integral and an application of Weierstrass' theorem imply that $x(t) \, \overline{\psi}(t) = 0$ a.e. on E. Relation (6.16.2) and the definition of ψ in (6.14.7) then imply that $x(t) = 0$ a.e. Hence $\mathfrak{H}_a(A) = L^2(E)$ and the proof is complete.

§ 6.17 Other singular integrals

Some of the previous results extend to more general types of integral operators. As above, let E denote a set of real numbers of positive measure and let $k(s, t)$ be a complex-valued, measurable, essentially bounded function on $E \times E$ satisfying $k(s, t) = \overline{k}(t, s)$ for which the integral operator K defined by

$$Kx = \pi^{-1} \int_E k(s, t) x(s) ds \qquad (6.17.1)$$

is semi-definite, that is, either $(Kx, x) \geq 0$ or $(Kx, x) \leq 0$ for all $x \in \mathfrak{D}_K$. If, as before, h is real-valued, measurable and essentially bounded on E, then define $A(h, k, E)$, analogous to $A(h, \phi, E)$ of (6.11.2), by

$$Ax = A(h, k, E)x = h(t)x(t) + (i\pi)^{-1} \int_E k(s, t)(s-t)^{-1}x(s)\,ds\,.$$

$$(6.17.2)$$

As above, in case E is bounded, then $AB - BA = -iK$. Whether or not E is bounded, but if E satisfies (6.16.1) for some interval I, the bounded operator B_μ of (6.14.5) can be introduced as before, and $AB_\mu - B_\mu A = iK_\mu$, where

$$K_\mu x = \pi^{-1} \int_E k(s, t)(t-\mu)^{-1}(s-\mu)^{-1} x(s)\,ds$$

is semi-definite. In either case Theorem 2.2.4 can be applied.

As an example one has the following result.

Theorem 6.17.1. *Suppose that $h(t), k(s, t)$ are defined as above and suppose that for any measurable subset E' of E,*

$$k(s, t) = 0 \text{ a.e. on } E \times E' \Rightarrow \text{meas } E' = 0\,. \qquad (6.17.3)$$

Then $A(h, k, E)$ of (6.17.2) is absolutely continuous.

Proof. First suppose that E is bounded. It follows from Theorem 2.2.4 that $\mathfrak{H}_a(A)$ contains all elements

$$\xi_{mn} = t^m \int_E k(s, t)s^n ds \text{ for } m, n = 0, 1, 2, \dots\,.$$

Hence if $x \in L^2(E)$ and $(x, \xi_{mn}) = 0$ it readily follows that $x(t)k(t, s) = 0$ a.e. on $E \times E$. That $x(t) = 0$ a.e. on E then follows from (6.17.3).

In case E is not bounded the argument is similar if B_μ is used rather than B.

§ 6.18 Reducing spaces of $A(0, \phi, E)$

Let $A = A(0, \phi, E)$ of (6.11.2), with $h \equiv 0$ and ϕ defined by (6.11.1), have the spectral resolution

$$A = A(0, \phi, E) = \int \lambda\,dE_\lambda\,. \qquad (6.18.1)$$

Suppose that (6.16.1) holds for some interval I and that μ belongs to the interior of I. Then define $\psi = \psi_\mu(t)$ by (6.14.7) and consider the space S_ψ, where

$$S_\psi = L^2((-\infty, \infty), d\sigma_\psi), \quad \sigma_\psi(\lambda) = \|E_\lambda\psi\|^2\,. \qquad (6.18.2)$$

Clearly, for any function $g \in S_\psi$, $g(A)\psi \in L^2(E)$ and satisfies

$$\|g(A)\psi\|^2 = \int_{-\infty}^{\infty} |g(\lambda)|^2 d\sigma_\psi(\lambda) \quad \left(= \int_{sp(A)} |g(\lambda)|^2 d\sigma_\psi(\lambda)\right). \qquad (6.18.3)$$

It is clear that the subspace

$$\Omega_\psi = \text{c.l.m. } \{g(A)\psi\} \tag{6.18.4}$$

reduces A and that if $x \in \Omega_\psi$ and corresponds to $g(\lambda) \in S_\psi$, so that $x = g(A)\psi$, then Ax corresponds to $\lambda g(\lambda)$. Thus, on the space S_ψ, one obtains a spectral representation for A when restricted to Ω_ψ. Since $A^n\psi = \phi Q^n \xi$, where

$$\xi(t) = (t-\mu)^{-1} \; (\in L^2(E)) \quad \text{and} \quad Qx = (i\pi)^{-1} \int_E |\phi(s)|^2 (s-t)^{-1} x(s) ds \tag{6.18.5}$$

it is clear that

$$\Omega_\psi = \text{c.l.m. } \{\phi Q^n \xi\} . \tag{6.18.6}$$

In the special case when $\phi \in L^2(E)$ (as, for example, is the case if E is bounded) and even if (6.16.1) is not assumed, consider the space S_ϕ where

$$S_\phi = L^2((-\infty, \infty), d\sigma_\phi), \quad \sigma_\phi(\lambda) = \|E_\lambda \phi\|^2 . \tag{6.18.7}$$

Corresponding to the earlier results, $g(A)\phi$ is defined whenever $g(\lambda) \in S_\phi$, and

$$\|g(A)\phi\|^2 = \int_{-\infty}^{\infty} |g(\lambda)|^2 \, d\sigma_\phi(\lambda) \quad \left(= \int_{sp(A)} |g(\lambda)|^2 \, d\sigma_\phi(\lambda) \right) . \tag{6.18.8}$$

Hence, if Ω_ϕ is the subspace

$$\Omega_\phi = \text{c.l.m. } \{g(A)\phi\} , \tag{6.18.9}$$

then Ω_ϕ reduces A and on the space S_ϕ one obtains a spectral representation of A when restricted to Ω_ϕ. It is clear that

$$\Omega_\phi = \text{c.l.m. } \{\phi Q^n 1_E\}, \tag{6.18.10}$$

where 1_E is the characteristic function of E.

§ 6.19 Estimates for σ_ψ and σ_ϕ

Suppose now that (6.16.2) holds and that E satisfies (6.16.1). If μ is in the interior of I, then (6.14.6) holds. Multiplications on the left and right of this equation by the projection $E(\Delta)$, where Δ is an arbitrary interval, lead to

$$\int_\Delta \lambda \, dE_\lambda B_{\mu\Delta} - B_{\mu\Delta} \int_\Delta \lambda \, dE_\lambda = iC_{\mu\Delta} ,$$

where

$$B_{\mu\Delta} = E(\Delta) B_\mu E(\Delta) \quad \text{and} \quad C_{\mu\Delta} = E(\Delta) C_\mu E(\Delta) .$$

Clearly, $C_{\mu\Delta} \geqq 0$ and $\|C_{\mu\Delta}\| = \pi^{-1} \|E(\Delta)\psi\|^2$; cf. § 6.14. An application of Theorem 2.2.1′, with the roles of the Hilbert space $\mathfrak{H}$ and the operators A, B taken now by $E(\Delta)\mathfrak{H}$, $B_{\mu\Delta}$ and $\int_\Delta \lambda \, dE_\lambda$ respectively, yields

$$\|E(\Delta)\psi\|^2 \leqq \tfrac{1}{2} |\Delta| \text{ meas sp}(B_{\mu\Delta}) , \tag{6.19.1}$$

where $|\varDelta|$ is the length of $\varDelta$. In view of the hypothesis (6.16.1), $A=A(0, \phi, E)$ is absolutely continuous (Theorem 6.16.1) and $\sigma_\psi(\lambda)$ is absolutely continuous. Hence relation (6.19.1) implies that for almost all λ,

$$d\sigma_\psi/d\lambda \le \tfrac{1}{2} \lim \inf \left(\text{meas sp} \left(B_{\mu\varDelta} \right) \right), \qquad (6.19.2)$$

where $\varDelta = (\lambda, \lambda + \varDelta\lambda)$ and $\varDelta\lambda \to 0$.

In case the set E is bounded the operator B of (6.12.2) is bounded and self-adjoint on $L^2(E)$. In addition, relation (6.12.3) holds, and, corresponding to (6.19.2), one now has that σ_ϕ is absolutely continuous and, for almost all λ,

$$d\sigma_\phi/d\lambda \le \tfrac{1}{2} \lim \inf \left(\text{meas sp} \left(E(\varDelta) \, B E(\varDelta) \right) \right), \qquad (6.19.3)$$

where $\varDelta = (\lambda, \lambda + \varDelta\lambda)$, $\varDelta\lambda \to 0$.

In case E is a half-line, say $E = (a, \infty)$ or a finite interval $E = (a, b)$ and ϕ satisfies

$$|\phi(t)| \equiv 1, \qquad (6.19.4)$$

then (6.19.2) and (6.19.3) can easily be refined to equalities. For if $E = (a, \infty)$, the "lim inf" expression of (6.19.2) is majorized by

$$\max \text{sp} \left(B_\mu \right) - \min \text{sp} \left(B_\mu \right) = (a - \mu)^{-1},$$

and so $d\sigma_\psi/d\lambda \le \tfrac{1}{2}(a-\mu)^{-1}$. Hence

$$(a-\mu)^{-1} = \int_a^\infty (t-\mu)^{-2} dt = \|\psi\|^2 \le \tfrac{1}{2}(a-\mu)^{-1} \text{ meas sp} (A) \le (a-\mu)^{-1}$$

and so, in this case,

$$d\sigma_\psi/d\lambda \equiv \tfrac{1}{2}(a-\lambda)^{-1}. \qquad (6.19.5)$$

If $E = (a, b)$ then the "lim inf" expression of (6.19.3) is majorized by $b - a$ and hence $d\sigma_\phi/d\lambda \le \tfrac{1}{2}(b-a)$. But

$$b - a = \|\phi\|^2 \le \tfrac{1}{2}(b-a) \text{ meas sp} (A) = b - a$$

and hence

$$d\sigma_\phi/d\lambda \equiv \tfrac{1}{2}(b-a). \qquad (6.19.6)$$

§ 6.20 Spectral representation for $A(0, 1, (a, b))$

It was shown by Koppelman and Pincus [2] that $A(0, 1, (a, \infty))$ is unitarily equivalent to $A(0, 1, (a, b))$. Also they obtained a spectral representation for this latter operator. This last result is given in the next theorem.

Theorem 6.20.1. *A spectral representation for* $A = A(0, 1, (a, b))$ *on* $L^2(a, b)$, *where* $-\infty < a < b < \infty$, *is obtained on the space* $L^2((-1, 1)$,

$\frac{1}{2}(b-a)\mathrm{d}\lambda)$. *In fact, for any $x \in L^2(a, b)$ there exists a function (essentially unique) $g(\lambda) \in L^2((-1, 1), \frac{1}{2}(b-a)\mathrm{d}\lambda)$ such that $x = g(A)\phi$ and*

$$\|x\|^2 = \tfrac{1}{2}(b-a) \int_{-1}^{1} |g(\lambda)|^2 \, \mathrm{d}\lambda \,. \tag{6.20.1}$$

Proof. As noted above, one proof can be found in Koppelman and Pincus [2]. Another is given in Putnam [31] and will be outlined below. Use will be made of the results of §§ 6.18, 6.19. In view of (6.18.10) it is sufficient to show that the functions $\{Q^n 1\}$, $n = 0, 1, 2, \ldots$, where

$$Qx = (i\pi)^{-1} \int_{a}^{b} (s-t)^{-1} x(s) \, \mathrm{d}s, \tag{6.20.2}$$

span $L^2(a, b)$. In order to do this it is enough to show that, for each $n = 0, 1, 2, \ldots$,

$$(Q1)^n = \sum_{k=0}^{n} a_k (Q^k 1), \quad a_n \neq 0, \tag{6.20.3}$$

and that

$$\text{c.l.m. } \{(Q1)^n\} = L^2(a, b) \,. \tag{6.20.4}$$

The relation (6.20.3) is established by some integral calculations, similar to those given in Koppelman [1], pp. 52 ff., and involving limits of powers of

$$F(z) = \int_{a}^{b} (s-z)^{-1} \, \mathrm{d}s \text{ for } \mathrm{Re}(z) \in [a, b] \text{ as } \mathrm{Im}(z) \to 0+ \text{ or } 0- \,.$$

Relation (6.20.4) is proved by showing that $(x, F^n) = 0$ for $n = 0, 1, 2, \ldots$, implies that $x = 0$ a.e. A change of variable shows that this amounts to showing that the powers of $\log [(b-t)/(t-a)]$ form a complete set on $L^2(a, b)$, which, in turn, can be established as a consequence of the uniqueness theorem for Fourier-Stieltjes transforms.

Remark. The operator $A(0, 1, (a, b))$ occurs in airfoil theory. Cf. Muskhelishvili [1], also Koppelman and Pincus [2] and the references cited there.

§ 6.21 Remarks on the spectra of singular integral operators

Schwartz [4] has defined the concept of essential spectrum for a general bounded operator T on a Hilbert space and has investigated its properties when T is a certain singular integral operator subject to differentiability assumptions on the coefficients. His methods use results from the theory of commutative Banach algebras. Schwartz also obtains the spectral resolution of a certain singular integral operator on $L^2(-\infty, \infty)$, using a variation of Friedrich's perturbation theory (Friedrichs [1, 2]).

Koppelman [2] and Pincus [1] have independently succeeded in diagonalizing the self-adjoint singular integral operator of the type (6.11.2) or similar to this, with E a finite interval, at least under certain, probably not completely essential, differentiability restrictions. Pincus [2] has extended his investigations to obtain a diagonalization of the self-adjoint operator J considered in § 3.12. Thus he determines the invariant spaces of the operator J satisfying $HJ-JH=(i/\pi)P$, where H is also self-adjoint, P is a one-dimensional projection, and the least space reducing H and containing the range of P is the entire Hilbert space $\mathfrak{H}$. The possibility of further generalization is also indicated.

§ 6.22 Jacobi matrices and absolute continuity

Let $A=(a_{ij})$ be the infinite matrix defined for $i, j=1, 2, \ldots$ by $a_{i+1,i}=b_i$ and $a_{ij}=0$ if $i \neq j+1$, where $|b_i| <$ const. Then A is a bounded matrix on

$$l^2 = \{x = (x_1, x_2, \ldots) : \sum_{i=1}^{\infty} |x_i|^2 < \infty\}$$

and it is easily verified that $A^*A - AA^* = D$ is a diagonal matrix with diagonal elements

$$|b_1|^2, \quad |b_2|^2 - |b_1|^2, \quad |b_3|^2 - |b_2|^2, \ldots .$$

If $A = H + iJ$ is the Cartesian representation of A then $H = \frac{1}{2}(A + A^*)$ is a special Jacobi matrix with zero diagonal elements. (See Wintner [2], p. 236. For an extensive discussion of general Jacobi matrices, see Stone [1].)

If, in addition, the b_i satisfy

$$|b_1| \leq |b_2| \leq \ldots . \tag{6.22.1}$$

then $D \geq 0$ and A is semi-normal, so that in particular, the previously obtained results on semi-normal operators are applicable. For instance, it follows from Theorem 3.2.1 that $\mathfrak{H}_a(H)$ (and also $\mathfrak{H}_a(J)$) each contains the least subspace of $\mathfrak{H}$ $(=l^2)$ reducing the matrix A (that is, invariant under A and A^*) and containing the range of D. Thus, H is surely absolutely continuous if

$$0 < |b_1| < |b_2| < \ldots \quad (\leq \text{const.}) . \tag{6.22.2}$$

That H may be absolutely continuous however without (6.22.2) being valid is easily seen; for instance, if all $b_n = 1$, H is the Jacobi matrix belonging to the quadratic form $\sum_{n=1}^{\infty} x_n x_{n+1}$ considered in § 6.8.

Whether H (or J) is absolutely continuous if (6.22.2) is weakened to, say, (6.22.1) and $A \neq 0$ (that is, $\Sigma|b_i|^2 > 0$) is apparently not known. Further questions concerning the nature of $\text{sp}(A)$ also seem to be open. For instance, although (3.5.16) holds with $T = A$, is also (3.5.17) valid?

Bibliography

Akhiezer, N. I. and I. M. Glazman:
  [1] *Theory of Linear Operators in Hilbert Space*, Frederick Ungar Pub. Co., N.Y., vol. I (1961), vol. II (1963).
Albert, A. A. and B. Muckenhoupt:
  [1] On matrices of trace zero, *Mich. Math. Jour.* **4**, 1–3 (1957).
Andô, T.:
  [1] On hyponormal operators, *Proc. Amer. Math. Soc.* **14**, 290–291 (1963).
Aronszajn, N.:
  [1] On a problem of Weyl in the theory of singular Sturm-Liouville equations, *Amer. Jour. Math.* **79**, 597–610 (1957).
Aronszajn, N. and W. F. Donoghue, Jr.:
  [1] On exponential representations of analytic functions in the upper halfplane with positive imaginary part, *Jour. d'Analyse Math.* **5**, 321–388 (1956–57).
Baker, H. F.:
  [1] Alternants and continuous groups, *Proc. London Math. Soc.* **3**, 24–47 (1905).
Beck, W. A. and C. R. Putnam:
  [1] A note on normal operators and their adjoints, *London Math. Soc.* **31**, 213–216 (1956).
Beckenbach, E. F. and R. Bellman:
  [1] Inequalities, Ergebnisse der Math. und ihrer Grenzgebiete, *New Series* **30**, (1961).
Bellman, R.
  [1] *Introduction to Matrix Analysis*, McGraw-Hill, N.Y. (1960).
Bendat, J. and S. Sherman:
  [1] Monotone and convex operator functions, *Trans. Amer. Math. Soc.* **79**, 58–71 (1935).
Berberian, S. K.:
  [1] Note on a theorem of Fuglede and Putnam, *Proc. Amer. Math. Soc.* **10**, 175–182 (1959).
  [2] *Introduction to Hilbert space*, Oxford Univ. Press, N. Y. (1961).
  [3] A note on hyponormal operators, *Pac. Jour. Math.* **12**, 1171–1175 (1962).
  [4] A note on operators unitarily equivalent to their adjoints, *Jour. London Math. Soc.*, **37**, 403–404 (1962).
  [5] The numerical range of a normal operator, *Duke Math. Jour.* **31**, 479–484 (1964).
Beurling, A.:
  [1] On two problems concerning linear transformations in Hilbert space, *Acta Math.* **81**, 239–255 (1949).
Birkhoff, G.:
  [1] Analytical groups, *Trans Amer. Math. Soc.* **43**, 61–101 (1938).
Birman, M.S.:
  [1] Perturbation of the spectrum of a singular elliptic operator under variation of the

boundary and boundary conditions, *Dokl. Akad. Nauk SSSR* **137**, 761–763 (1961) (in Russian); translation, *Sov. Math. Dokl.* **2**, 326–328 (1961).

[2] Conditions for the existence of wave operators, *Dokl. Akad. Nauk SSSR*, **143**, 506–509 (1962) (in Russian); translation, *Sov. Math. Dokl.* **3**, 408–411 (1962).

[3] A test for the existence of wave operators, *Sov. Dokl. Akad. Nauk SSSR*, **147**, boundary and boundary conditions, *Dokl. Akad. Nauk SSSR* **137**, 761–763 (1961) 1008–1009 (1962) (in Russian); translation, *Sov. Math. Dokl.* **3**, 1747–1748 (1962).

[4] Existence conditions for wave operators, *Izv. Akad. Nauk SSSR Ser. Mat.* **27**, 883–906 (1963).

[5] A local criterion for the existence of wave operators, *Dokl. Akad. Nauk SSSR* **159**, 485–488 (1964) (in Russian); translation, *Sov. Math. Dokl.* **5**, 1505–1509 (1964).

Birman, M.S. and S. B. Entina:

[1] A stationary approach in the abstract theory of scattering, *Dokl. Akad. Nauk SSSR* **155**, 506–508 (1964) (in Russian); translation, *Sov. Math. Dokl.* **5**, 432–435 (1964).

Birman, M.S. and M.G. Krein:

[1] On the theory of wave operators and scattering operators, *Dokl. Akad. Nauk SSSR* **144**, 475–478 (1962) (in Russian); translation, *Sov. Math. Dokl.* **3**, 740–747 (1962).

Bishop, E.:

[1] Spectral theory for operators on a Banach space, *Trans. Amer. Math. Soc.* **86**, 414–445 (1957).

Bogoliubov, N.N. and D.V. Shirkov:

[1] *Introduction to the Theory of Quantized Fields*, Interscience Pub., Inc., N.Y. (1959).

de Branges, L.:

[1] Perturbations of self-adjoint transformations, *Amer. Jour. Math.* **84**, 543–560 (1962).

Brown, A.:

[1] On a class of operators, *Proc. Amer. Math. Soc.* **4**, 723–728 (1953).

Brown, A. and P. R. Halmos:

[1] Algebraic properties of Toeplitz operators, *Jour. für die reine und angewandte Math.* **213**, 89–102 (1963).

Brown, A., P. R. Halmos and C. Pearcy:

[1] Commutators of operators on Hilbert space, *Can. Jour. Math.* **17**, 695–708 (1965).

Brown, A. and C. Pearcy:

[1] Structure theorem for commutators of operators, *Bull. Amer. Math. Soc.* **70**, 779–780 (1964).

[2] Structure of commutators of operators, *Annals of Math.* **82**, 112–127 (1965).

[3] Multiplicative commutators of operators, *Can. Jour. Math.* **18**, 737–749 (1966).

Brownell, F. H.:

[1] A note on Cook's wave-matrix theorem, *Pac. Jour. Math.* **12**, 47–52 (1962).

Calderon, A. P.:

[1] Singular integrals, *Bull. Amer. Math. Soc.* **72**, 427–465 (1966).

Calderon, A. P. and A. Zygmund:

[1] Singular integral operators and differential equations, *Amer. Jour. Math.* **79**, 901–921 (1957).

Calkin, J. W.:

[1] Two-sided ideals and congruences in the ring of bounded operators in Hilbert space, *Annals of Math.* **42**, 839–873 (1941).

Campbell, J. E.:

[1] On a law of combination of operators bearing on the theory of continuous transformation groups, *Proc. London Math. Soc.* **28**, 381–390 (1897).

Cesari, L.:
[1] *Asymptotic Behavior and Stability Problems in Ordinary Differential Equations*, Springer-Verlag (1959).

Coddington, E. A. and N. Levinson:
[1] *Theory of Ordinary Differential Equations*, McGraw-Hill Book Co., Inc., N.Y., Toronto, London (1955).

Cook, J. M.:
[1] Convergence to the Møller wave-matrix, *Jour. Math. Phys.* **36**, 82–87 (1957).

Cordes, H. O.:
[1] Über die Spektralzerlegung von hypermaximalen Operatoren, die durch Separation der variablen zerfallen, II Mitteilung, *Math. Ann.* **128**, 373–411 (1955).

Deckard, D. and Pearcy, C.:
[1] Another class of operators without square roots, *Proc. Amer. Math. Soc.* **14**, 445–449 (1963).

Devinatz, A.:
[1] Toeplitz operators on H^2 spaces, *Trans. Amer. Math. Soc.* **112**, 304–317 (1964).

Dixmier, J.:
[1] Sur la relation $i(PQ-QP)=1$, *Comp. Math.* **13**, 263–270 (1958).

Dollard, J. C.:
[1] Asymptotic convergence and the Coulomb interaction, *Jour. Math. Phys.* **5**, 729–738 (1964).

Donoghue, W. F.:
[1] On the numerical range of a bounded operator, *Mich. Math. Jour.* **4**, 261–263 (1957).
[2] On a problem of Nieminen, *Inst. Hautes Etudes Sci. Publ. Math.*, No. 16, 31–33 (1963).

Döring, W.:
[1] *Einführung in die Quantenmechanik*, Göttingen (1962).

Douglas, R. G. and C. Pearcy:
[1] Spectral theory of generalized Toeplitz operators, *Trans. Amer. Math. Soc.* **115**, 433–444 (1965).

Dunford, N.:
[1] Spectral Theory II. Resolutions of the identity, *Pac. Jour. Math.* **2**, 559–614 (1952).
[2] Spectral operators, *Pac. Jour. Math.* **4**, 321–354 (1954).
[3] A survey of the theory of spectral operators, *Bull. Amer. Math. Soc.* **64**, 217–274 (1958).

Dunford, N. and J. Schwartz:
[1] *Linear Operators*, Interscience, N.Y., vol. 1 (1958); vol. 2 (1964).

Faddeev, L. D.:
[1] Mathematical problems of the quantum theory of scattering for a three-particle system, *Trudy Mat. Inst. Steklov* **69** (1963) (in Russian); translation, United Kingdom Atomic Energy Authority, Library, Atomic Energy Research Establishment, Harwell, Berkshire (1964).

Foiaş, C. and L. Gehér:
[1] Über die Weylsche Vertauschungsrelationen, *Acta Sci. Math.* (*Szeged*) **24**, 97–102 (1963).

Foiaş, C., L. Gehér and B. Sz.-Nagy:
[1] On the permutability condition of quantum mechanics, *Acta Sci. Math.* (*Szeged*) **21**, 78–89 (1960).

Freeman, J. M.:
[1] The perturbation theory of some Volterra operators, *Thesis, M.I.T.* June 1963.

Friedrichs, K. O.:
[1] Über die Spektralzerlegung eines Integraloperators, *Math. Ann.* **115**, 249–272 (1938).
[2] On the perturbation of continuous spectra, *Comm. App. Math.* **1**, 361–406 (1948).
[3] *Mathematical Aspects of the Quantum Theory of Fields*, Interscience Pub., N.Y. (1953).
Frobenius, G.:
[1] Series of papers appearing in Sitzgsber. Preuss. Akad. Wiss., 1896 et seq.
Fuglede, B.:
[1] A commutativity theorem for normal operators, *Proc. Nat. Acad. Sci.* **36**, 35–40 (1950).
Gårding, L. and A. Wightman:
[1] Representations of the anticommutation relations, *Proc. Nat. Acad. Sci.* **40**, 617–621 (1954).
[2] Representations of the commutation relations, *Proc. Nat. Acad. Sci.* **40**, 622–626 (1954).
Gelfand, I. M. and B. M. Levitan:
[1] Determination of a differential equation by its spectral function, *Izv. Akad. Nauk SSSR* **15**, 309–360 (1951) (in Russian).
Gonshor, H.:
[1] Spectral theory for a class of non-normal operators, *Can. Jour. Math.* **8**, 449–461 (1956).
Green, T. A. and O. E. Lanford, III, Rigorous derivation of the phase shift formula for the Hilbert space scattering operator of a single particle, *Jour. Math. Phys.* **1**, 139–148 (1960).
Grenander, U. and G. Szegö:
[1] *Toeplitz Forms and Their Applications*, Univ. of Calif. (1958).
Hack, M. N.:
[1] On convergence to the Møller wave operators, *Nuovo Cim.* **9**, 731–733 (1958).
Halmos, P. R.:
[1] Commutation and spectral properties of normal operators, *Acta Sci. Math. (Szeged)* **12**, 153–156 (1950).
[2] Normal dilations and extensions of operators, *Summa Brasil. Math.* **2**, 125–134 (1950).
[3] *Introduction to Hilbert Space and the Theory of Spectral Multiplicity*, Chelsea Pub. Co., N.Y. (1951).
[4] Commutators of operators, *Amer. Jour. Math.* **74**, 237–240 (1952).
[5] Commutators of operators, II, *Amer. Jour. Math.* **76**, 191–198 (1954).
[6] *Lectures on Ergodic Theory*, The Math. Soc. of Japan (1956).
[7] *Finite-Dimensional Vector Spaces*, van Nostrand Co., N.Y. (1958),
[8] Shifts on Hilbert Spaces, *Jour. für die reine und angewandte Math.* **208**, 102–112 (1961).
[9] A glimpse into Hilbert space, *Lectures on Mathematics* (Chapt. I), Wiley, N.Y. (1963).
[10] Numerical ranges and normal dilations, *Acta Sci. Math. (Szeged)* **25**, 1–5 (1964).
Halmos, P. R. and G. Lumer:
[1] Square roots of operators II, *Proc. Amer. Math. Soc.* **5**, 589–595 (1954).
Halmos, P. R., G. Lumer and J. J. Schäffer:
[1] Square roots of operators, *Proc. Amer. Math. Soc.* **4**, 142–149 (1953).
Hardy, G. H., J. E. Littlewood and G. Pólya:
[1] *Inequalities*, Cambridge (1934).
Hartman, P.:
[1] On the essential spectra of symmetric operators in Hilbert space, *Amer. Jour. Math.* **75**, 229–240 (1953).

[2] Aurel Wintner, *Jour. London Math. Soc.* **37**, 483–503 (1962).

[3] On unbounded Toeplitz matrices, *Amer. Jour. Math.* **85**, 59–78 (1963).

Hartman, P. and A. Wintner:

[1] On the spectra of Toeplitz's matrices, *Amer. Jour. Math.* **72**, 359–366 (1950).

[2] The spectra of Toeplitz's matrices, *Amer. Jour. Math.* **76**, 867–882 (1954).

Hausdorff, F.:

[1] Die symbolische Exponentialformel in der Gruppentheorie, *Leipzig Ber.* **58**, 19–48 (1906).

Heinz, E.:

[1] Beiträge zur Störungstheorie der Spektralzerlegung, *Math. Ann.* **123**, 415–438 (1951).

Hellinger, E. D.:

[1] Spectra of quadratic forms in infinitely many variables, *Part IV of Mathematical Monographs*, Northwestern Univ. **1** (1941).

Herstein, I. N.:

[1] On a theorem of Putnam and Wintner, *Proc. Amer. Math. Soc.* **8**, 535–536 (1957).

Hilbert, D.:

[1] *Grundzüge einer allgemeinentheorie der linearen Integralgleichungen*, Leipzig (1912).

Hildebrandt, S.:

[1] The closure of the numerical range of an operator as spectral set, *Comm. Pure and App. Math.* **17**, 415–421 (1964).

[2] Numerischer Wertebereich und normale Dilatationen, *Acta Sci. Math. (Szeged)* **26**, 187–190 (1965).

[3] Über den numerischen Wertebereich eines Operators, *Math. Ann.* **163**, 230–247 (1966).

Hilgevoord, J.:

[1] *Dispersion Relations and Causal Description*, North-Holland Pub. Co., Amsterdam (1960).

Hill, C. K.:

[1] On the singly-infinite Hilbert matrix, *Jour. London Math. Soc.* **35**, 17–29 (1960).

Hille, E.:

[1] On roots and logarithms of elements of a complex Banach algebra, *Math. Ann.* **136**, 46–57 (1958).

Hille, E. and R. S. Phillips:

[1] Functional analysis and semi-groups, *Amer. Math. Soc.*, Colloquium Pub. **31** (1957).

Hopf, E.:

[1] *Ergodentheorie*, Chelsea Pub. Co., N.Y. (1948).

Hörmander, L.:

[1] *Linear Partial Differential Operators*, Academic Press, N.Y. (1963).

Ikebe, T.:

[1] Eigenfunction expansions associated with the Schrödinger operators and their applications to scattering theory, *Arch. Rat. Mech. and Anal.* **5**, 1–34 (1960).

Ionescu-Tulcea, A.:

[1] Random series and spectra of measure-preserving transformations, *Proc. Symposium on Ergodic Theory*, Tulane Univ. (1961).

Ismagilov, R.S.:

[1] The spectrum of Toeplitz matrices, *Dokl. Akad. Nauk SSSR* **149**, 769–772 (1963) (in Russian); translation, *Sov. Math. Dokl.* **4**, 462–465 (1963).

Jacobson, N.:

[1] Rational methods in the theory of Lie algebras, *Annals of Math.* **36**, 875–881 (1935).

Jauch, J. M.:

[1] Theory of the scattering operator, *Helv. Phys. Acta* **31**, 127–158 (1958).

Jauch, J. M. and I. I. Zinnes:
[1] The asymptotic condition for simple scattering systems, *Il Nuovo Cim.* **11**, 553–567 (1959).
Jordan, P. and E. Wigner:
[1] Über das Paulische Äquivalenzverbot, *Zeits. für Phys.* **47**, 631–651 (1928).
Kaplansky, I.:
[1] Products of normal operators, *Duke Math. Jour.* **20**, 257–260 (1953).
[2] Functional analysis–part of book: *Some Aspects of Analysis and Probability*, Surveys in App. Math., John Wiley & Sons, Inc., N.Y. (1958).
Kato, T.:
[1] Fundamental properties of Hamiltonian operators of Schrödinger type, *Trans. Amer. Math. Soc.* **70**, 195–211 (1951).
[2] On finite-dimensional perturbations of self-adjoint operators, *Jour. Math. Soc. Japan* **9**, 239–249 (1957).
[3] Perturbation of continuous spectra by trace class operators, *Proc. Japan Acad.* **33**, 260–264 (1957).
[4] Scattering operators and perturbation of continuous spectra, *Sûgaku* **9**, 75–84 (1957) (in Japanese).
[5] On the Hilbert matrix, *Proc. Amer. Math. Soc.* **8**, 73–80 (1957).
[6] On the commutation relation $AB - BA = C$, *Arch. for Rat. Mech and Anal.* **10**, 273–275 (1963).
[7] Wave operators and unitary equivalence, Tech. Rep. Univ. of Calif. (1963); *Pac. Jour. Math.* **15**, 171–180 (1965).
[8] Wave operators and similarity for some non-selfadjoint operators, *Math. Ann.* **162**, 258–279 (1966).
Kato, T. and S. T. Kuroda:
[1] A remark on the unitarity property of the scattering operator, *Nuovo Cim.* **14**, 1102–1107 (1959).
Kato, T. and O. Taussky:
[1] Commutators of A and A^*, *Jour. Wash. Acad. Sci.* **46**, 38–40 (1956).
Kellogg, O. D.:
[1] *Foundations of Potential Theory*, Springer, Berlin (1929).
Kilpi, Y.:
[1] Über das komplexe Momentenproblem, *Ann. Acad. Sci. Fenn.* Ser. A, I. Math., 236 (1957).
[2] Zur Theorie der Quantenmechanischen Vertauschungsrelationen, *Ann. Acad. Sci. Fenn.*, Ser. A, I. Math., 315 (1962).
Kleinecke, D. C.:
[1] On operator commutators, *Proc. Amer. Math. Soc.* **8**, 535–536 (1957).
Kodaira, K.:
[1] The eigenvalue problem for ordinary differential equations of the second order and Heisenberg's theory of S-matrices, *Amer. Jour. Math.* **71**, 921–945 (1949).
Koppelman, W.:
[1] On the spectral theory of singular integral operators, *Trans. Amer. Math. Soc.* **97**, 35–63 (1960).
[2] Spectral multiplicity theory for a class of singular integral operators, *Trans. Amer. Math. Soc.* **113**, 87–100 (1964).
Koppelman, W. and J. D. Pincus:
[1] Perturbations of continuous spectra and singular integral operators, *Report No.* 242, New York Univ., N.Y. (1957).
[2] Spectral representation for finite Hilbert transformations, *Math. Zeits.* **71**, 399–407 (1959).

Krabbe, G. L.:
[1] On the logarithm of a uniformly bounded operator, *Trans Amer. Math. Soc.* **81**, 155–166 (1956).

Kramers, H. A.:
[1] *The Foundations of Quantum Theory*, North-Holland Pub. Co., Amsterdam, Interscience, N.Y. (1957).

Krein, M. G.:
[1] On the trace formula in perturbation theory, *Mat. Sbor.* **33**, (75), 597–626 (1953) (in Russian).
[2] On a method of effective solution of an inverse boundary problem, *Dokl. Akad. Nauk SSSR* **94**, 987–990 (1954) (in Russian).
[3] Perturbation determinants and a formula for the traces of unitary and self-adjoint operators, *Dokl. Akad. Nauk SSSR* **144**, 268–271 (1962) (in Russian); translation, Sov. Math. Dokl. **3**, 707–710 (1962).
[4] *On Some New Investigations in the Perturbation Theory of Self-adjoint Operators*, Kiev, 103–187 (1963) (in Russian).

Kreith, K. and F. Wolf:
[1] On the effect on the essential spectrum of the change of the basic region, *Indag. Math.* **22**, 312–315 (1960).

Kristensen, P., L. Mejlbo and E. T. Poulsen:
[1] Tempered distributions in infinitely many dimensions, I. Canonical field operators, *Comm. Math. Phys.* **1**, 175–214 (1965).
[2] Tempered distributions in infinitely many dimensions, II. Displacement operators, *Math. Scand.* **14**, 129–150 (1964).

Kurepa, S.:
[1] On normal n-th roots of a self-adjoint operator, *Glasnik Mat.-Fiz. Astr.* Zagreb **3**, 163–169 (1960).
[2] On n-th roots of normal operators, *Math. Zeits.* **78**, 285–292 (1962).
[3] On operator-roots of an analytic function, *Glasnik Math-Fiz. Astr.*, 49–51 (1963).

Kuroda, S. T.:
[1] On a theorem of Weyl-von Neumann, *Proc. Japan Acad.* **34**, 11–15 (1958).
[2] On the existence and unitary property of the scattering operator, *Nuovo Cim* **12**, 431–454 (1959).
[3] Perturbation of continuous spectra by unbounded operators, I, *Jour. Math. Soc. Japan* **11**, 247–262 (1959).
[4] Perturbation of continuous spectra by unbounded operators, II, *ibid.* **12**, 243–257 (1960).
[5] On a paper of Green and Lanford, *Jour. Math. Phys.* **3**, 933–935 (1962).
[6] Finite-dimensional perturbation and a representation of scattering operator, *Pac. Jour. Math.* **13**, 1305–1318 (1963).
[7] On a stationary approach to scattering problem, *Bull. Amer. Math. Soc.* **70**, 556–560 (1964).

Kuzmin, E. N.:
[1] On commutators in elastic rings, *Sbir. Mate. Zh.* **1** No. 2, 198–204 (1960) (in Russian).

Löwner, K.:
[1] Über monotone Matrixfunktionen, *Math. Zeits.* **38**, 177–216 (1934).

Ludwig, G.:
[1] *Die Grundlagen der Quantenmechanik*, Springer, Berlin, Göttingen, Heidelberg (1954).

Lumer, G. and M. Rosenblum:
[1] Linear operator equations, *Proc. Amer. Math. Soc.* **10**, 32–41 (1959).

Mackey, G. W.:
[1] *Mathematical Foundations of Quantum Mechanics*, Math. Phys. Monograph Ser. W. A. Benjamin, Inc., N.Y. (1963).

154   Bibliography

McCoy, N. H.:
  [1] On quasi-commutative matrices, *Trans. Amer. Math. Soc.* **36**, 327–340 (1934).
Magnus, W.:
  [1] On the spectrum of Hilbert's matrix, *Amer. Jour. Math.* **72**, 699–704 (1950).
Marcus, M. and N. A. Khan:
  [1] On matrix commutators, *Can. Jour. Math.* **12**, 269–277 (1960).
Mejlbo, L. C.:
  [1] On the solution of the commutation relation $PQ - QP = -iI$, *Math. Scand.* **13**, 129–139 (1963).
Meng, Ching-Hwa:
  [1] On the numerical range of an operator, *Proc. Amer. Math. Soc.* **14**, 167–171 (1963).
Møller, C.:
  [1] General properties of the characteristic matrix in the theory of elementary particles, I, *Kgl. Dan. Videnskab. Selskab. Math.-Fys. Medd.* **23**, (1945).
Murray, F. J. and J. von Neumann:
  [1] On rings of operators: I, *Ann. Math.* **37**, 116–229 (1936).
Muskhelishvili, N. I.:
  [1] *Singular Integral Equations*, Noordhoff, Groningen, Holland (1953).
Naimark, M. A.:
  [1] *Normed Rings*,Noordhoff, Groningen (1954).
von Neumann, J.:
  [1] Allgemeine Eigenwerttheorie Hermitescher Funktionaloperatoren, *Math. Ann.* **102**, 49–131 (1929).
  [2] Die Eindeutigkeit der Schrödingerschen Operatoren, *Math. Ann.* **104**, 570–578 (1931).
  [3] Über adjungierte Funktionaloperatoren, *Ann. of Math.* **33**, 294–310 (1932).
  [4] On infinite direct products, *Comp. Math.* **6**, 1–77 (1938).
  [5] Charakterisierung des Spektrums eines Integraloperators, *Act. Sci. et Ind.* **229** (1935).
  [6] Approximative properties of matrices of high finite order. *Port. Math.* **3**, 1–62 (1942).
  [7] *Mathematical Foundations of Quantum Mechanics*, Princeton Univ. Press, Princeton (1955).
Nieminen, T.:
  [1] A condition for the self-adjointness of a linear operator, *Ann. Acad. Sci. Fenn.*, Ser. A, I. Math., 316 (1962).
Orland, G. H.:
  [1] On a class of operators, *Proc. Amer. Math. Soc.* **15**, 75–79 (1964).
Pasiencier, S. and H.-C. Wang:
  [1] Commutators in a semi-simple Lie group, *Proc. Amer. Math. Soc.* **13**, 907–913 (1962).
Pearcy, C.:
  [1] On commutators of operators on Hilbert space, *Proc. Amer. Math. Soc.* **16**, 53–59 (1965).
Peter, F. and H. Weyl:
  [1] Die Vollständigkeit der primitiven Darstellungen einer geschlossenen kontinu-ierlichen Gruppe, *Math. Ann.* **97**, 737–755 (1927).
Pincus, J. D.:
  [1] On the spectral theory of singular integral operators, *Trans. Amer. Math. Soc.* **113**, 101–128 (1964).
  [2] *Commutators, Generalized Eigenfunction Expansions and Singular Integral Operators*, Brookhaven Nat. Lab. AMD 363 (1964); *Trans. Amer. Math. Soc.* **121**, 358–377 (1966).

Povzner, A. J.:
[1] On the expansion of arbitrary functions in characteristic functions of the operator
   $-\Delta u + cu$, *Mat. Sbor. N.S.* **32** (74), 109–156 (1953) (in Russian).
Prosser, R. T.:
[1] Convergent perturbation expansions for certain wave operators, *Jour. Math. Phys.* **5**, 708–713 (1964).
Putnam, C. R.:
[1] On commutators of bounded matrices, *Amer. Jour. Math.* **73**, 127–131 (1951).
[2] On normal operators in Hilbert space, *Amer. Jour. Math.* **73**, 357–362 (1951).
[3] The quantum mechanical equations of motion and commutation relations, *Phys. Rev.* **74**, 1047–1048 (1951).
[4] The spectra of quantum mechanical operators, *Amer. Jour. Math.* **74**, 377–388 (1952).
[5] On function space-Hilbert space correspondences in quantum mechanics, *Quart. Jour. Math.* (Oxford) **3**, 260–267 (1952).
[6] On the continuous spectra of singular boundary value problems, *Can. Jour. Math.* **6**, 420–426 (1954).
[7] Remarks on certain operators of quantum field theory, *Jour. London Math. Soc.* **29**, 350–356 (1954).
[8] On the spectra of commutators, *Proc. Amer. Math. Soc.* **5**, 929–931 (1954).
[9] On commutators and Jacobi matrices, *Proc. Amer. Math. Soc.* **7**, 1026–1030 (1956).
[10] Continuous spectra and unitary equivalence, *Pac. Jour. Math.* **7**, 993–995 (1957).
[11] On semi-normal operators, *Pac. Jour. Math.* **7**, 1649–1652 (1957).
[12] On square roots of normal operators, *Proc. Amer. Math. Soc.* **8**, 768–769 (1957).
[13] Commutators and absolutely continuous operators, *Trans. Amer. Math. Soc.* **87**, 513–525 (1958).
[14] Commutators and normal operators, *Port. Math.* **17** Fasc. 2, 59–62 (1958).
[15] On the numerical ranges of commutators, *Jour. London Math. Soc.* **34**, 23–26 (1959).
[16] On Toeplitz matrices, absolute continuity, and unitary equivalence, *Pac. Jour. Math.* **9**, 837–846 (1959).
[17] On differences of unitarily equivalent self-adjoint operators, *Proc. Glas. Math. Assoc.* **4**, 103–107 (1960).
[18] A note on Toeplitz matrices and unitary equivalence, *Boll. della Unione Mat. Ital.* **15**, 6–9 (1960).
[19] Group commutators of bounded operators in Hilbert space, *Mich. Math. Jour.* **7**, 229–232 (1960).
[20] On the spectra of group commutators, *Boll. della Unione Mat. Ital.* **15**, 379–383 (1960).
[21] Commutators, perturbations, and unitary spectra, *Acta Math.* **106**, 215–232 (1961).
[22] On the spectra of unitary half-scattering operators, *Quart. App. Math.* **20**, 85–88 (1962).
[23] Absolute continuity of certain unitary and half-scattering operators, *Proc. Amer. Math. Soc.* **13**, 844–846 (1962).
[24] The spectra of irrotational flows, *Quart. App. Math.* **20**, 388–389 (1963).
[25] On semi-normal operators and convex hulls of ranges of power series, *Jour. London Math. Soc.* **38**, 218–222 (1963).
[26] Absolutely continuous Hamiltonian operators, *Jour. Math. Anal. and Appl.* **7**, 163–165 (1963).
[27] On the structure of semi-normal operators, *Bull. Amer. Math. Soc.* **69**, 818–819 (1963).

[28] Toeplitz matrices and invertibility of Hankel matrices, *Pac. Jour. Math.* **14**, 651–658 (1964).

[29] Commutators, absolutely continuous spectra, and singular integral operators, *Amer. Jour. Math.* **86**, 310–316 (1964).

[30] Wave operators and absolutely continuous spectra, *Quart. App. Math.* **22**, 256–260 (1964).

[31] The spectra of generalized Hilbert transforms, *Jour. Math. and Mech.* **14**, 857–872 (1965).

[32] On the spectra of semi-normal operators, *Trans. Amer. Math. Soc.* **119**, 509–523 (1965).

Putnam, C. R. and A. Wintner:

[1] The orthogonal group in Hilbert space, *Amer. Jour. Math.* **74**, 52–78 (1952).

[2] On the spectra of group commutators, *Proc. Amer. Math. Soc.* **9**, 360–362 (1958).

Rejto, P. A.:

[1] On gentle perturbations, I. *Comm. Pure and Appl. Math.* **16**, 279–303 (1963).

[2] On gentle perturbations, II. *Comm. Pure and Appl. Math.* **17**, 257–292 (1964).

[3] Some absolutely continuous operators, *N.Y.U. Courant Inst. Math. Sci.* IMM-NYU 329 (1964).

[4] Some absolutely continuous operators II, Math. Research Center, U.S. Army, *Tech. Summary Report* 582, July 1965.

Rellich, F.:

[1] Störungstheorie der Spektralzerlegung, III. Mitt., *Math. Ann.* **116**, 555–570 (1939).

[2] Der Eindeutigkeitssatz für die Lösungen der quantenmechanischen Vertauschungs-relationen, *Nachr. Akad. Wiss. Gött., Math.-Phys. Klasse*, 107–115 (1946).

[3] Halbbeschränkte gewöhnliche Differentialoperatoren zweiter Ordnung, *Math. Ann.* **122**, 343–368 (1950).

Riesz, F.:

[1] *Les systèmes d'équations linéaires à une infinité d'inconnues*, Paris (1913).

Riesz, F. and M. Riesz:

[1] Über die Randwerte einer analytischen Funktion, *Quat. Cong. des Math. Scand.* Stockholm, 27–44 (1916).

Riesz, F. and B. Sz.-Nagy:

[1] *Functional Analysis*, Frederick Ungar Pub. Co., N.Y. (1955).

Robinson, D. W.:

[1] A note on matrix commutators, *Mich. Math. Jour.* **7**, 31–33 (1960).

Rosenblum, M.:

[1] On the operator equation $BX - XA = Q$, *Duke Math. Jour.* **23**, 263–270 (1956).

[2] Perturbation of the continuous spectrum and unitary equivalence, *Pac. Jour. Math.* **7**, 997–1010 (1957).

[3] On the Hilbert matrix (I), *Proc. Amer. Math. Soc.* **9**, 137–140 (1958).

[4] On the Hilbert matrix (II), *Proc. Amer. Math. Soc.*, **9**, 581–585 (1958)

[5] On a theorem of Fuglede and Putnam, *Jour. London Math. Soc.* **33**, 376–377 (1958).

[6] The absolute continuity of Toeplitz matrices, *Pac. Jour. Math.* **10**, 987–996 (1960).

[7] Self-adjoint Toeplitz operators and associated orthonormal functions, *Proc. Amer. Math. Soc.* **13**, 590–595 (1962).

[8] A concrete spectral theory for self-adjoint Toeplitz operators, *Amer. Jour. Math.* **87**, 709–718 (1965).

Rutherford, D. E.:

[1] On the solution of the matrix equation $AX + XB = C$, *Akad. van Weten.*, Amsterdam, Proc. **35**, 54–59 (1932).

Saitô, T. and T. Yoshino:

[1] Note on the canonical decomposition of contraction, *Tôhoku Math. Jour.* **16**, 309–312 (1964).

[2] On a conjecture of Berberian, *Tôhoku Math. Jour.* **17**, 147–149 (1965).

Sakai, S.:

[1] On some problems of C^*-algebras, *Tôhoku Math. Jour.* **11**, 453–455 (1959).

[2] On a conjecture of Kaplansky, *Tôhoku Math. Jour.* **12**, 31–33 (1960).

Sarason, D. E.:

[1] *The Theorem of F. and M. Riesz on the Absolute Continuity of Analytic Measures*, Thesis, Univ. of Mich. (1963).

Schäffer, J. J.:

[1] On some problems concerning operators in Hilbert space, *Anais Acad. Brasil. Ci.* **25**, 87–90 (1953).

[2] On unitary dilations of contraction operators, *Proc. Amer. Math. Soc.* **6**, 322 (1955).

Schatten, R.:

[1] *A Theory of Cross Spaces*, Ann. Math. Studies, Princeton, (1950).

[2] *Norm Ideals of Completely Continuous Operators*, Springer, Berlin (1960).

Schreiber, M.:

[1] Unitary dilations of operators, *Duke Math. Jour.* **23**, 579–594 (1956).

[2] A functional calculus for general operators in Hilbert space, *Trans. Amer. Math. Soc.* **87**, 108–118 (1958).

[3] On the spectrum of a contraction, *Proc. Amer. Math. Soc.* **12**, 709–713 (1961).

[4] Absolutely continuous operators, *Duke Math. Jour.* **29**, 175–190 (1962).

[5] Numerical range and spectral sets, *Mich. Math. Jour.* **10**, 283–288 (1963).

Schwartz, J.:

[1] Some non-selfadjoint operators, *Comm. Pure and Appl. Math.* **13**, 609–639 (1960).

[2] Some non-selfadjoint operators, II. A family of operators yielding to Friedrich's method, *Comm. Pure and Appl. Math.* **14**, 619–626 (1961).

[3] A remark on inequalities of Calderon-Zygmund type for vector-valued functions, *Comm. Pure and Appl. Math.* **14**, 785–799 (1961).

[4] Some results on the spectra and spectral resolutions of a class of singular integral operators, *Comm. Pure and Appl. Math.* **15**, 75–90 (1962).

Segal, I.E.:

[1] *Mathematical Problems of Relativistic Physics*, Lectures in Appl. Math., vol. II, Amer. Math. Soc., Providence, R.I. (1963).

Shinbrot, M.:

[1] On singular operators, *Jour. Math. and Mech.* **13**, 395–406 (1964).

Shoda, K.:

[1] Einige Sätze über Matrizen, *Jap. Jour. Math.* **13**, 361–365 (1936).

[2] Über den Kommutator der Matrizen, *Jour. Math. Soc. Japan* **3**, 78–81 (1951).

Singer, I. M. and J. Wermer:

[1] Derivations on commutative normed algebras, *Math. Ann.* **129**, 260–264 (1955).

Sirokov, F. V.:

[1] Proof of a conjecture of Kaplansky, *Usp. Math. Nauk* **11**, 4 (70), 167–168 (1956) (in Russian).

Stampfli, J. G.:

[1] Hyponormal operators, *Pac. Jour. Math.* **12**, 1453–1458 (1962).

[2] Roots of scalar operators, *Proc. Amer. Math. Soc.* **13**, 796–798 (1962).

[3] Hyponormal operators and spectral density, *Trans. Amer. Math. Soc.* **117**, 469–476 (1965).

[4] Remarks on monotone shifts, Univ. of Mich., ONR, NR 043–319, March, 1965.

[5] A study of the Dunford boundedness condition for certain classes of subnormal and hyponormal operators, Univ. of Mich., ONR, NR 043–319, May, 1965.

[6] Extreme points of the numerical range of a hyponormal operator, *Mich. Math. Jour.* **13**, 87–89 (1966).

[7] Minimal range theorems for operators with thin spectra, preprint.

Stankevič, I. V.:
[1] On the theory of perturbations of a continuous spectrum, *Dokl. Akad. Nauk SSSR* **144**, 279–282 (1962) (in Russian); *Sov. Math. Dokl.* **3**, 719–722 (1962).

Stone, M. H.:
[1] *Linear Transformations in Hilbert Space and Their Applications to Analysis*, Amer. Math. Soc., N.Y. (1932).

Streater, R. F. and A. S. Wightman:
[1] *PCT, Spin and Statistics, and All That*, W. A. Benjamin, N.Y. and Amsterdam (1964).

Stummel, F.:
[1] Singuläre elliptische Differentialoperatoren in Hilbertschen Räumen, *Math. Ann.* **132**, 150–176 (1956).

Sz.-Nagy, B.:
[1] *Spektraldarstellung linearer Transformationen des Hilbertschen Raumes*, Berlin, (1942).
[2] On uniformly bounded linear transformations in Hilbert space, *Acta Sci. Math. (Szeged)* **11**, 152–157 (1947).
[3] Sur les contractions de l'espace de Hilbert, *Acta Sci. Math. (Szeged)* **15**, 87–92 (1953).
[4] Transformations de l'espace de Hilbert, fonctions de type positif sur un groupe, *Acta Sci. Math. (Szeged)* **15**, 104–114 (1954).
[5] Sur les contractions de l'espace de Hilbert, II, *Acta Sci. Math. (Szeged)* **18**, 1–14 (1957).
[6] *Extensions of Linear Transformations in Hilbert Space which Extend beyond this Space*, Appendix to Riesz and Sz.-Nagy *Functional Analysis*, Frederick Ungar Pub. Co., N.Y. (1960).

Sz.-Nagy, B. and C. Foiaş:
[1] Sur les contractions de l'espace de Hilbert, III, *Acta Sci. Math. (Szeged)* **19**, 26–46 (1958).
[2] Sur les contractions de l'espace de Hilbert, IV, *Acta Sci. Math. (Szeged)* **21**, 251–259 (1960).

Taussky, O.:
[1] A note on the group commutator of A and A^*, *Jour. Wash. Acad. Sci.* **48**, 305 (1958).
[2] Commutators of unitary matrices which commute with one factor, *Jour. Math. and Mech.* **10**, 175–178 (1961).

Taylor, A. E.:
[1] Spectral theory of closed distributive operators, *Acta Math.* **84**, 189–224 (1951).
[2] *Introduction to Functional Analysis*, John Wiley & Sons, Inc., N.Y. (1958).

Thompson, R. C.:
[1] On matrix commutators, *Jour. Wash. Acad. Sci.* **48**, 306–307 (1958).

Tillmann, H. G.:
[1] Zur Eindeutigkeit der Lösungen der quantenmechanischen Vertauschungsrelationen, *Acta Sci. Math. (Szeged)* **24**, 258–270 (1963).
[2] Zur Eindeutigkeit der Lösungen der quantenmechanischen Vertauschungsrelationen, II, *Arch. der Math.* **15**, 332–334 (1964).

Titchmarsh, E. C.:
[1] *Eigenfunction Expansions Associated with Second-order Differential Equations*, Oxford (1946).
[2] *Eigenfunction Expansions Associated with Second-order Differential Equations*, part II, Oxford (1958).
[3] *Introduction to the Theory of Fourier Integrals*, Oxford (1948).

Toeplitz, O.:
[1] Zur Theorie der quadratischen Formen von unendlichvielen Veränderlichen, *Gött. Nachr.*, 489–506 (1910).
[2] Zur Theorie der quadratischen und bilinearen Formen von unendlichvielen Veränderlichen, *Math. Ann.* **70**, 351–376 (1911).
[3] Über die Fouriersche Entwicklung positiver Funktionen, *Rend. Circ. Mat. Pal.* **32**, 191–192 (1911).

Venkataraman, V. K.:
[1] *On Commutators and Absolutely Continuous Operators in a Hilbert Space*, Thesis, Purdue Univ. (1963).

Vidav, I.:
[1] Über eine Vermutung von Kaplansky, *Math. Zeits.* **62**, 330 (1955).

Weyl, H.:
[1] Über beschränkte quadratische Formen deren Differenz vollstetig ist, *Rend. Circ. Math. Pal.* **27**, 373–392 (1909).
[2] Über gewöhnliche Differentialgleichungen mit Singularitäten und die zugehörigen Entwicklungen willkürlicher Funktionen, *Math. Ann.* **68**, 220–269 (1910).
[3] Quantenmechanik und Gruppentheorie, *Zeits. für Phys.* **46**, 1–47 (1928).

Widom, H.:
[1] Singular integral equations in L_p, *Trans. Amer. Math. Soc.* **97**, 131–160 (1960).
[2] On the spectrum of a Toeplitz operator, *Pac. Jour. Math.* **14**, 365–375 (1964).

Wiegmann, N. A.:
[1] A note on infinite normal matrices, *Duke Math. Jour.* **16**, 535–538 (1949).

Wielandt, H.:
[1] Über die Unbeschränkheit der Schrödingerschen Operatoren der Quantenmechanik, *Math. Ann.* **121**, 21 (1949).

Wiener, N.:
[1] *The Fourier Integral*, Cambridge, (1933).

Wiener, N. and A. Wintner:
[1] On the ergodic dynamics of almost periodic systems, *Amer. Jour. Math.* **63**, 794–824 (1941).

Wienholtz, E.:
[1] Halbbeschränkte partielle Differentialoperatoren zweiter Ordnung von elliptischen Typus, *Math. Ann.* **135**, 50–80 (1958).

Wigner, E. P.:
[1] Do the equations of motion determine the quantum mechanical commutation relations?, *Phys. Rev.* **77**, 711–712 (1950).

Wigner, E. P. and M. M. Yanase:
[1] On the positive semidefinite nature of a certain matrix expression, *Can. Jour. Math.* **16**, 397–406 (1964).

Wintner, A.:
[1] Zur Theorie der beschränkten Bilinearformen, *Math. Zeits.* **30**, 228–282 (1929).
[2] *Spektraltheorie der unendlichen Matrizen*, Leipzig (1929).
[3] On non-singular bounded matrices, *Amer. Jour. Math.* **54**, 145–149 (1932).
[4] On normal inertia, *Studia Math.* **8**, 135–140 (1939).
[5] Small perturbations, *Amer. Jour. Math.* **67**, 417–430 (1945).
[6] The unboundedness of quantum-mechanical matrices, *Phys. Rev.* **71**, 738–739 (1947).
[7] *The Analytical Foundations of Celestial Mechanics*, Princeton Univ. Press, Princeton, (1947).
[8] On the logarithms of bounded matrices, *Amer. Jour. Math.* **74**, 360–364 (1952).

Wolf, F.:
  [1] Perturbation by changes of one-dimensional boundary conditions, *Indag. Math.*
      18, 360–366 (1956).
  [2] On the essential spectrum of partial differential boundary problems, *Comm. Pure
      and Appl. Math.* **12**, 211–228 (1959).
  [3] On the invariance of the essential spectrum under a change of boundary conditions
      of partial differential boundary operators, *Indag. Math.* **21**, 142–147 (1959).
  [4] On the perturbation of an elliptic operator which leaves the essential spectrum
      invariant, *Acad. Roy. Belg. Bull. Cl. Sci.* **46**, 441–447 (1960).
Xa Dao-xeng (Hsia Tao-hsing):
  [1] On non-normal operators, *Chinese Math.* **3**, 232–246 (1963).
Yoshino, T.:
  [1] On the spectrum of a hyponormal operator, *Tôhoku Math. Jour.* **17**, 305–309 (1965).
Zaanen, A. C.:
  [1] *Linear Analysis*, North-Holland Pub. Co., Amsterdam (1953).

Symbol Index

(Numbers denote pages where symbols are defined or first used)

Author Index

Subject Index